T0136747

extraordinary
orchids

Sandra Knapp

With a Foreword by Mark Chase

The University of Chicago Press

The University of Chicago Press,
Chicago 60637

Published 2021
Printed in China

30 29 28 27 26 25 24 23 22 21 1 2 3 4 5

ISBN-13: 978-0-226-77967-6 (cloth)
ISBN-13: 978-0-226-77970-6 (e-book)
DOI: https://doi.org/10.7208/
chicago/9780226779706.001.0001

First published in 2021 by the Natural History
Museum, Cromwell Road, London SW7 5BD.

Library of Congress Cataloging-in-Publication Data

Names: Knapp, Sandra, author.
Title: Extraordinary orchids / Sandra Knapp.
Description: Chicago : The University of Chicago Press,
2021. | "First published in 2021 by the Natural History
Museum, . . . London."—Title page verso. | Includes
bibliographical references and indexes.
Identifiers: LCCN 2020029317 | ISBN 9780226779676 (cloth) |
ISBN 9780226779706 (ebook)
Subjects: LCSH: Orchids.
Classification: LCC QK495.O64 K58 2021 | DDC 584/.72—dc23
LC record available at https://lccn.loc.gov/2020029317

Designed by Bobby Birchall, Bobby&Co Design
Reproduction by Saxon Digital Services, UK

Printed by Toppan Leefung Printing Limited

Contents

Foreword

I am one of those people who was passionate about orchids before I even became a botanist and, in fact, was drawn to botany by my fascination with them. Initially I studied the history of East Asia and grew African violets (*Streptocarpus*) as a hobby, but became bored with them and started experimenting with more exotic members of that family Gesneriaceae. Then I bought my first orchid, a *Cattleya* hybrid, and when it flowered just a few weeks later, my *Koehleria*, *Achimines* and *Columnea* plants, beautiful as they were, had to go. My orchid collection grew swiftly and I was compelled to switch to study biology so I could research orchids and understand why they are so unusual. So began my lifelong entrancement.

My focus was on orchid taxonomy and my PhD on a small, mostly Mexican genus, *Leochilus*. At that time in 1985, it had just become possible to carry out DNA studies, in an emerging field of specialization called molecular systematics. This was a perfect vehicle to answer so many questions: what are the major groups of orchids, how are these groups related to each other, when did orchids evolve, and what other families of plants are they related to. When I started my work none of these questions could be satisfactorily answered. A lot has changed. Part of that major change is in the number of orchid scientists. In the 1980s, a worldwide meeting of orchid scientists would have numbered less than 20, but now we have at least two orders of magnitude more. Many serious scientists had been discouraged from orchid studies by the sheer size of the family and the lack of a scientific and phylogenetic framework on which to base their research. Those who did study orchids were often characterized as fanatics and dilettantes. Now those attitudes have changed, as evidenced by this book.

What we have learned about these extraordinary plants is remarkable. We now know there are five major groups, recognized taxonomically as subfamilies: Apostasioideae, Vanilloideae, Cypripedioideae, Orchidoideae and Epidendroideae, the last two much larger than the first three. Orchids are members of the large order Asparagales, which includes asparagus, daffodil, amaryllis, onion, agave, aloe, yucca, grass tree and daylily (but not the true lilies, tulips etc. which are members of the distantly related order Liliales). DNA data can be used as a 'molecular clock' to determine the time of origin for a group of plants without a fossil record. Previously, orchids were thought, due to the lack of a fossil record, to be recently evolved. However, by using this DNA-based clock, we have learned that orchids are one of the oldest families of plants and originated before the end of the Cretaceous, 66 million years ago. Early orchids grew throughout the then closely spaced land masses, and coexisted with dinosaurs, but they survived the great Cretaceous extinction and are now one of the two largest plant families. This co-existence with the dinosaurs makes me wonder if small dinosaurs might have pollinated these early orchids, given that modern orchids are pollinated by such a wide range of animals. Obviously, any orchids depending on a dinosaur for pollination would have died out with the dinosaurs, and since we have no orchid fossils from that time it is difficult to think how we could prove such a phenomenon ever existed.

The great number of orchid scientists today reflects what excellent subjects they are – extreme examples of many plant traits. As a general rule in biology, if you want to study a particular process select an extreme example because this makes it easier to tease apart the various factors involved. Orchids fit this model well and have become one of the groups of choice for studying pollination, physiology of photosynthesis, mycology, genomics, flower development, and many other topics.

Finally, as Sandy has noted, orchids are an important group on which to focus conservation efforts. They are the botanical 'canaries in the mine' and perfect examples of the complicated, interconnectedness of the natural world. They depend on fungi to enable their germination and establishment, trees to provide sites on which to grow, and a huge range of animals to pollinate them, most of which depend on other plant species for their livelihood because most orchids are floral cheats and offer their pollinators no reward. If wild orchids, our botanical 'canaries in the mine', are threatened by pollution, land clearance and climate change, then the rest of nature is also in deep trouble. To conserve orchids, we have to maintain ecosystems that can also support the myriad of other organisms upon which they depend, and, by studying orchids, we will be gaining insight into one of the most remarkable phenomena of the natural world. It is not at all surprising to me that, after he finished *On the Origin of Species*, Darwin's attention turned to orchids.

Mark Chase, *Royal Botanic Gardens, Kew*

Preface

For me orchids have always been like the loud, over-dressed guest at the party – over the top. They are one of the most species-rich families of flowering plants, only daisies perhaps outdo them, but then that is a matter of how you count species. It is an ongoing debate. An urban legend has it that one botanist said to another, 'You know, we all hate orchids, don't you?' I could have been that botanist; for many years I have been both fascinated and slightly repelled by orchids, but everyone else loves them, so perhaps they have enough admirers. Looking more deeply into their diversity and everyday lives while writing this book, however, has made me appreciate them more. Orchids are an exemplar of all that is fascinating about plants in general – how and where they grow and how they attract pollinators in order to reproduce, and their complex interactions with a host of other organisms are quite surprising. Plants are truly extraordinary, their apparent stillness and lack of behaviour conceals a great deal that is left to discover, if only we look carefully and through a lens that is not animal-centric or anthropomorphic. Plants truly support the planet; they warrant our attention, and ultimately our protection and support. Orchids, with their sometimes over-the-top flowers and extreme lifestyles, can open our eyes to other interactions in the plant world that may have escaped our notice. Their beauty and form draw our attention, and their rarity and often inaccessibility makes searching for orchids obsessive. Whether it be the anti-hero of Susan Orlean's book *The Orchid Thief*, or the vandals who stripped a local hillside for the rare *Phragmipedium besseae* where I once lived in Peru, people have extreme reactions to orchids. *National Geographic* photographer Carlton Wade, who photographed the rare ghost orchid *Dendrophylax lindenii* in the Florida Everglades, mused 'I do think it's possible that orchids drive people crazy.' That may be true, but love of orchids has also led to breakthroughs in plant conservation, and the protection of habitats where other species thrive as well. I may not have been driven crazy, but I do love orchids more now than I did before – I hope you will too, whether you start from love or disdain.

Sandra Knapp, *London*

OPPOSITE: The butterfly orchid, *Psychopsis papilio,* is widespread in South America. The large flowers are held on tall stalks that sway in the breeze looking for all the world like butterflies in flight, giving the plant its common name.

Life in the air

'We shall now describe the Aristocrats of wild plants,
who convey their nobility by wanting to live only
high up in the trees, and never below
on the ground, [...] wherefore they have a strange
way of growing and are strangely fashioned,
just like Aristocrats flaunting their finery.'

Georg Everhard Rumphius (1741–1750)
Het Amboinsche Kruit-boek (Ambonese Herbal)
translated from the Dutch by
Beekman in *Rumphius' Orchids,* 2003

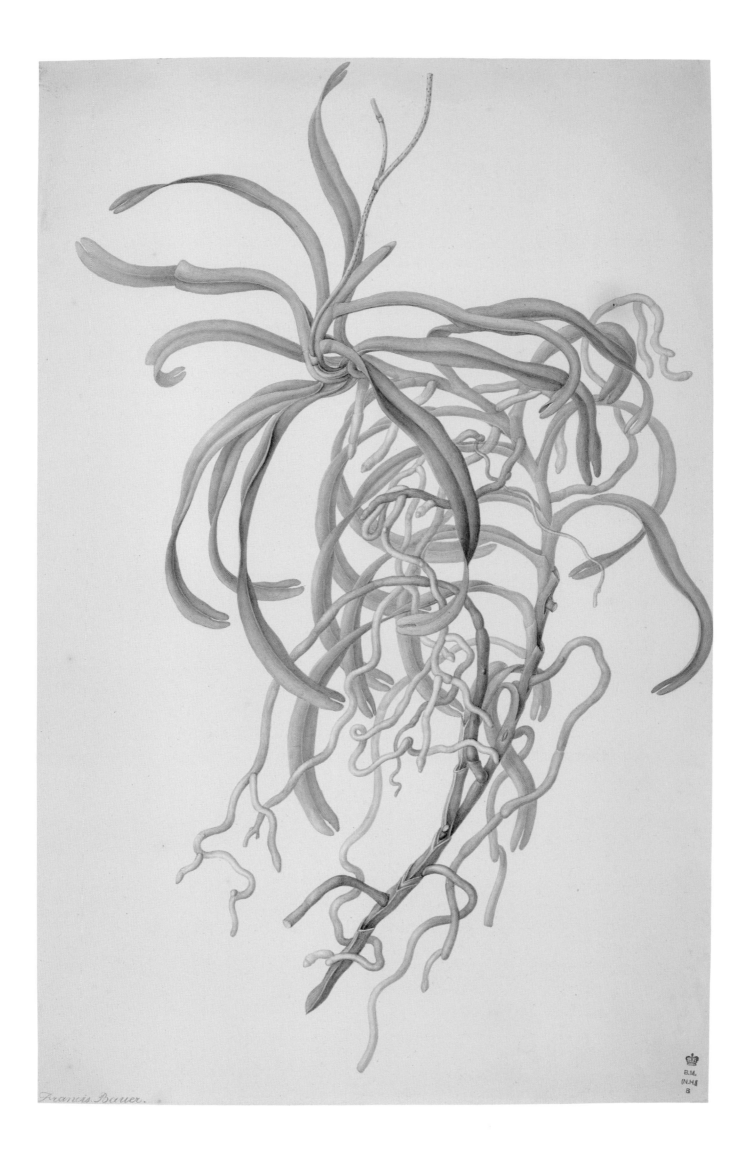

Francis Bauer.

THINK OF ORCHIDS, and jungles spring to mind – dense impenetrable forests that are hot, humid and dangerous, with branches festooned with colourful flowers. This is the image of the tropical rainforest the young Alfred Russel Wallace, co-discoverer of evolution by natural selection with Charles Darwin, carried with him to the Amazon, but he was sorely disappointed when he found that all he saw was green – the festoons of orchids were at the tops of the trees. Many, if not most, tropical orchids are epiphytic; that is, they live on other plants (the word comes from the Greek *epi*, meaning 'on', and *phyte*, meaning 'plant'). Early tropical explorers thought of these plants as parasites. Sir Hans Sloane, botanist and physician to the Duke of Albemarle who was Lieutenant Governor of Jamaica from 1687 to 1688, described them as a kind of mistletoe, naming one *Viscum radice bulbosa majus & elatius, delphinii flore ferrugineo & guttato* (a Neotropical orchid today known by the Linnaean bionomial *Maxillaria alba*) and characterizing it as growing 'on the Trunks and Arms of trees, as Mistletoe or others of this kind...'.

Today, *Viscum* is the scientific name for the mistletoe we see hanging in doorways at Christmas time, and mistletoes are true parasites that send root-like structures called haustoria into the host to tap into the nutrients being produced by the host plant. But orchids and other tropical epiphytes such as bromeliads (pineapple relatives), ferns and mosses are not parasites; they live perched on branches and do not depend upon their host for nutrition. Just like the trees themselves, they photosynthesize to take up carbon dioxide and release oxygen, making their own food in the form of sugars. Living in the air is the epiphytic lifestyle. About 10 per cent of flowering plant species are epiphytes, and orchids contribute hugely to this number; in fact, the preponderance of the epiphytic lifestyle may help to account for the high species diversity of the orchid family in general.

It seems like it would be ideal living in the canopy of the rainforest with plenty of sunlight, plenty of water – perfect for plant growth and success. But living in the canopy can be stressful indeed. High above the forest floor, temperature and water availability fluctuate wildly, with the extremes higher and lower than anywhere else in the forest, so epiphytes have to cope with daily peaks and troughs of sunlight, water and wind. Tree canopies themselves moderate the conditions below – filtering torrential rain, damping

OPPOSITE: The silvery white outer layer, or velamen, of epiphytic orchid roots, like this *Cleisostoma paniculatum*, is formed of cells that lose their contents as they mature – they are air-filled and act as sponges for water and nutrients in the forest canopy.

down wind gusts and providing shade, in effect creating the tropical rainforest environment we usually think of. No more than a fraction of the light available at the top makes it to the forest floor to be available for photosynthesis by the herbs and shrubs of the forest understorey. But up in the canopy, when the sun shines it is unrelenting, and when it rains there is no respite. Orchids deal with these challenges in a number of special ways.

First, the leaves. Epiphytic orchid leaves are usually thick, stiff and somewhat fleshy; some are even round in cross section. Many species also have swollen stems, some so inflated that they are called pseudobulbs, or 'false' bulbs, because they are not buried underground like those of tulips and daffodils, which are made up of overlapping leaf bases rather than being stems. These thick leaves and swollen stems help orchid plants retain water in times of drought – think of a cactus, the stem of which can store water for long periods, and then imagine these leaves and stems performing the same function, but high in the canopy.

Plants need water not only to grow, but also to perform photosynthesis, the reaction in which water and carbon dioxide are converted to sugars and oxygen, much to our benefit! The basic reaction of photosynthesis requires three main things: water, carbon dioxide and sunlight. The energy in sunlight drives the whole process. Water is usually taken up by plant roots, and all plants take in atmospheric gases such as carbon dioxide through tiny pores in the leaves called stomates. The problem with these pores is they also cause the plant to lose water via evaporation, especially if it is hot, as in the middle of the day. Plants can lose around 97 per cent of the water they absorb through their roots to evaporation. How can this be efficient in an environment where water can be limiting? As a result most, if not all, epiphytic orchids use a special form of photosynthesis called crassulacean acid metabolism (CAM), which compartmentalizes the reactions so that the stomata can be closed during the day when water loss is a risk, and open at night to take up carbon dioxide. CAM is a photosynthetic pathway that was first discovered in succulents of the family Crassulaceae, which includes the jade plant and sedums, and is also found in many other plant families living in stressful environments. In CAM photosynthesis, carbon dioxide taken up through the stomata at night is converted to an acid and stored in vacuoles – hollow spaces in cells – during

the hours of darkness to be then transported to the chloroplast and reconverted to carbon dioxide for use in photosynthesis once the sun comes out. This specialized form of photosynthesis allows the plants to save water in environments where it is at a premium, including high in the forest canopy of the tropics.

Orchid roots are also specialized for living in the air. If you have ever looked at the roots of an epiphytic orchid, you might notice they are covered by a spongy white layer that looks almost like very thin foam. This white layer is called the velamen, and its structure greatly increases uptake of water and atmospheric gases, allowing the plant better access to both water and nutrients. The velamen is formed of layers of cells that die as the root matures. If you look at the very tip of an epiphytic orchid root, it is green; sometimes the velamen cells can harbour cyanobacteria – these nitrogen-fixing bacteria could be helping the plant by providing extra nutrition in the low nutrient environment of the forest canopy. Some of the roots of epiphytic orchids anchor the plant to tree trunks or branches, where they form part of an epiphytic community. And community it is – orchids are rarely the only ones up high in the canopy. Along with orchids, the branches of tropical trees bear communities of mosses, lichens and sometimes other epiphytic plants. These aggregations can create soil, and even supply the tree itself with minerals that diffuse from the soil through the bark into the host tree.

Epiphytes also create mini-ecosystems for insects and other arthropods, and certainly contribute to the extremely high species diversity of tropical forests. The most specialized of these ecosystems is known as an ant garden. Here, various species of ant build their nests among the tangled roots of epiphytes and sometimes even use

ABOVE: *Orchis mascula*. The opening and closing of stomates on leaf undersides are controlled by changes in turgor pressure in the crescent-shaped guard cells.

orchid stems themselves as nesting places; the hollow pseudobulbs of some orchids like *Schomburgkia* and *Myrmecophila* harbour colonies of ants, and the orchid itself takes up nutrients from the detritus the ants accumulate. Most tropical orchids are not obligate members of these ant garden ecosystems, but when they are, their tiny seeds are removed and moved about by the ants, potentially to new branches and growing sites. These complex mutualisms in tropical forests, where both partners in the relationship derive benefit, are common and becoming ever better understood as scientists are able to access the canopy in new ways to study how it works.

The sheer variety of form found in tropical orchids spurred a mania for orchid collection and cultivation in the nineteenth century. The glamour and rareness of these plants was the ultimate prize for gardeners wanting to show off their collections. Termed 'Orchido-mania' by James Bateman in his 1843 mega-book *The Orchidaceae of Mexico and Guatemala,* this new condition 'pervades all (and especially the upper) classes to such a marvellous extent.' That 'Orchido-mania' was especially prevalent in the 'upper classes' was due to the fact that cultivating tropical orchids required a considerable infrastructure. First, the glasshouses, then the heating, then the light. All this cost money, and a lot of it. But in the beginning, it did not go all the way of the aristocratic growers. Shipment upon shipment of tropical epiphytes came to Europe, and most either died on route or were potted up carefully in heavy, rich soils, only to die later on.

It was assumed that being from the tropics, orchids needed hot and steamy conditions, so in they went to over-heated humid glasshouses. It took a lot of experimentation to tease out the best conditions for cultivating these seemingly delicate plants. In the early 1800s, Joseph Banks – of the voyage of HMS *Endeavour* fame and then President of the Royal Society – invented a hanging basket filled with moss and twigs in which to grow his specimens; he was more successful than most.

The code of orchid growing was finally cracked when growers began to think carefully about where each of these wonderful plants was found. Some, yes, were from the hot, humid lowland tropics, but many of the new and glamorous plants coming into cultivation were from the higher mountains of the tropics – quite a different environment. High elevation epiphytic orchids suffer from heat more than from cold, something which seems counterintuitive but which growers began to master. Bateman

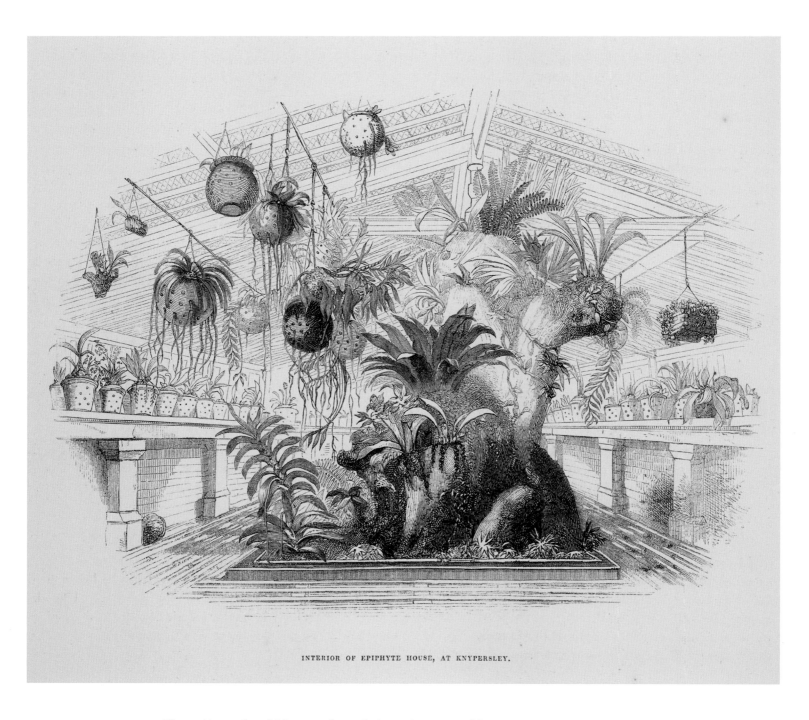

INTERIOR OF EPIPHYTE HOUSE, AT KNYPERSLEY.

ABOVE: The culture of orchids away from their native tropical forest habitats was revolutionized by the realization that they did not need constant water and heat. Bateman's sketch of the ideal orchid house was one stimulus for orchid cultivation for the masses, although he felt orchids and the aristocracy belonged together!

Pl. 33.

Mrs Withers del.   M. Gauci lith.

ONCIDIUM WENTWORTHIANUM.

Publ. by J. Ridgway & Sons, 169 Piccadilly. March, 1843.
Printed by P. Gauci.

pointed out that the orchids he was treating in his masterpiece were 'more abundant in the higher latitudes and purer air, than in the hot and pestiferous jungles of the coast' – so these tropical plants were different. He described his growing techniques in wonderfully flowery language, and a plan for an orchid conservatory completes the instruction. The rules were simple: '1st The plants can scarcely have too much light or too much sun', '2nd Take care of the roots', '3rd Beware of noxious insects', '4th Give the plants a season of rest', '5th Attend to the condition of the air' and '6th Do not over-water'. Pretty simple, really.

The publication of John Charles Lyons's manual, *Remarks on the Cultivation of Orchidaceous Plants*, published in 1843, the same year as Bateman's tome, gave clear instructions that didn't cost the earth – orchid cultivation no longer needed to be the preserve of the aristocracy with money to spare for glasshouses or skilled gardeners. In the middle part of the nineteenth century, orchid cultivation became something for (almost) everyone, sparking a revolution in cultivation. A series of articles entitled 'Orchids for the million' was published in the 1850s by a gardener called Benjamin Williams at the instigation of the great Victorian orchid taxonomist John Lindley. This manual provided even simpler instructions and eventually became *The Orchid Grower's Manual*, a set of instructions by which anyone could grow epiphytic orchids, even in their living rooms – something we still see today.

Victorian orchid fanciers preferred the large-flowered show-offs of the orchid world – as still seems the fashion today, given what is available for sale in greenhouses and supermarkets – but orchids in nature are more diverse than just those we see on the supermarket shelves. Orchids come in all shapes and sizes, from the cattleyas with their blousy flowers to *Lepanthes* or *Oberonia* with flowers one can barely see. Fashions in orchid cultivation in the nineteenth century favoured the showy and magnificent, and Victorian collectors satisfied this desire with more and more new species and genera from the tropics. But they were also collecting the less obvious species, sending these to specialists such as John Lindley in Cambridge, whose interest in orchids spanned the lot. So how did these collectors reach the canopy to collect these orchidaceous rarities? How does one access an ecosystem 30 metres (98 feet) or more in the air?

OPPOSITE: The long inflorescences of *Oncidium* species, like this *Oncidium wentworthianum*, dangle freely, looking almost like butterflies suspended in mid-air. Whilst on the Amazon Alfred Russel Wallace was enchanted with the sight, 'But what lovely yellow flower is that suspended in the air between two trunks, yet far from either? It shines from the gloom as if its petals were of gold. Now we pass it close by, and see its stalk, like a slender wire a yard and a half long, springing from a cluster of thick leaves on the bark of a tree. It is an *Oncidium*, one of the lovely orchis tribe, making these gloomy shades gay with its airy and brilliant flowers.'

Sometimes branches fall or trees are felled; when I collected plants in the American tropics I would follow the road builders and harvest the epiphytes from the fallen trees, making specimens for scientific study. That was highly opportunistic though, and not a specialized collection of orchids for cultivation. The collectors who supplied the Victorian enthusiasts often lived in tropical areas and would employ local people to climb, or cut down, trees to get at the plants high in the branches. Sometimes they even used trained monkeys to access the exotic 'air plants' that grew out of reach. I'm sure they, like me, also took advantage of fallen trees and branches. They probably observed plants *in situ* to see if the flowers were exciting enough to warrant sending to Europe, and maybe even cultivated them. These collectors and growers were focusing on individual plants as commodities, not attempting to understand how they fit into a functioning ecosystem or what they did in their everyday lives.

Real access to the canopy in order to study it as a functioning ecosystem came with the adaptation of mountain-climbing techniques for scaling trees safely, first in the old growth forests of the Pacific Northwest of the Americas, and later on in the tropical rainforests. Today, the canopy can be accessed by a variety of methods, from tree-climbing to canopy rafts and walkways. Scientists are able to study *in situ* how plants and animals adapt to their environment and just how many organisms live in the canopy itself. Using tree-fogging, entomologists estimate that each tropical tree contains thousands of species of insects – multiply that by the number of trees, and it becomes mind-boggling. By being at the tops of the trees themselves for more than a fleeting moment, scientists discovered the stresses associated with living in the canopy, and untangled the adaptations that epiphytes, including orchids, have evolved to deal with these challenges. It is extraordinary how many challenges have been overcome to bring us that supermarket orchid – by orchid growers over the generations and by the orchids themselves in their life above the ground.

OPPOSITE: Jean Jules Linden, a Belgian horticulturist and explorer in the mid-nineteenth century, studied orchids, such as this *Masdevallia wageneriana*, in their native habitats. His observations changed how orchid growers treated these high elevation cloud forest epiphytes.

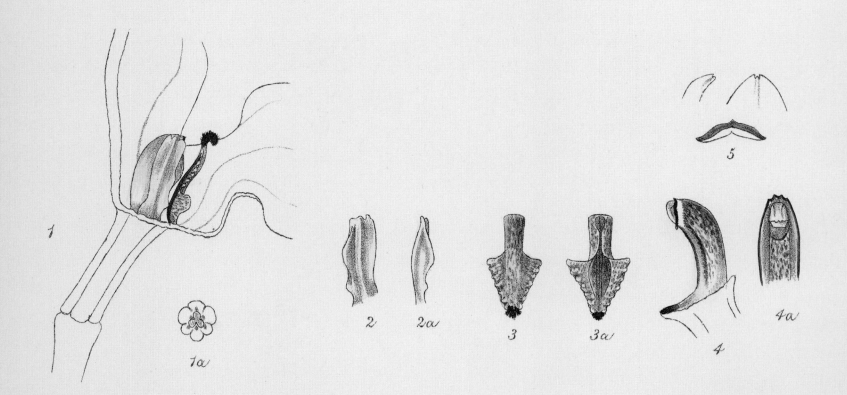

Plate VI.

W.H.Fitch, del.et lith.

Vincent Brooks, Imp.

Odontoglossum pendulum.

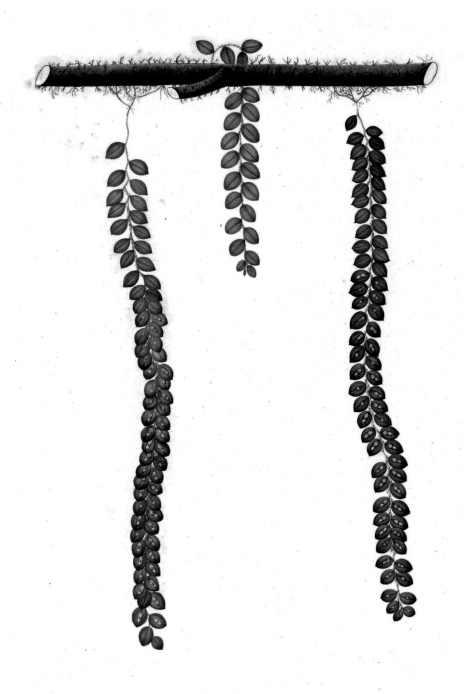

ABOVE: The size variation in orchids is immense, although all can be classified botanically as herbaceous. This tiny orchid, *Andinia nummularia*, with dangling stems and its tiny flowers less than a centimetre long tucked tightly against the leaves, hardly fits the blousy concept of the prom-night orchid at all.

OPPOSITE: This Mexican species, *Cuitlauzina pendula*, was placed into the genus *Odontoglossum* by James Bateman, but about 100 years later, the distinctness of *Cuitlauzina* was recognized, and the original name given to this plant became its correct name. The genus name honours Cuiltahuatzin (also known as Cuitláhuac), a brother of the Aztec ruler Moctezuma, whose rule was cut short by the Spanish invasion of Mexico.

ABOVE: John Gould was obsessed by hummingbirds, 'my thoughts are often directed to them in the day, and my night dreams have not infrequently carried me to their native forests'. The plates in his self-published paean to these wonders were done by his wife, Elizabeth, and show the birds as they would appear in nature. Here she depicts the Ecuadorian piedtail with an orchid for atmosphere; the hummingbird is not the pollinator of this *Paphinia*!

OPPOSITE: The tiny branches of tropical trees can support the tiniest of orchid plants like these miniature epiphytes. In 1830 the Victorian orchid specialist John Lindley named the genus *Oberonia* in honour of Oberon, Shakespeare's king of the fairies from the play *A Midsummer Night's Dream*. Three years later the Austrian botanist Stefan Endlicher named the same plants *Titania*, in honour of Oberon's wife. We use Lindley's name because it was published earlier – but Shakespeare provided inspiration in unexpected places twice over.

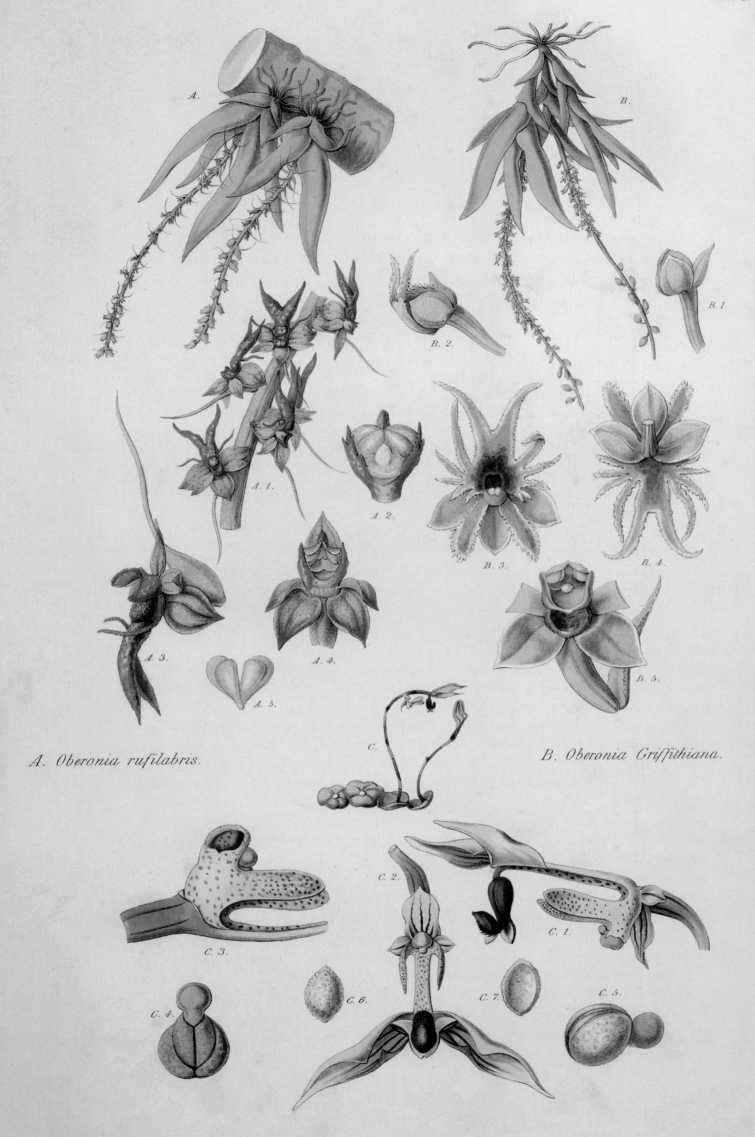

Pl. 8

A. Oberonia rufilabris.

B. Oberonia Griffithiana.

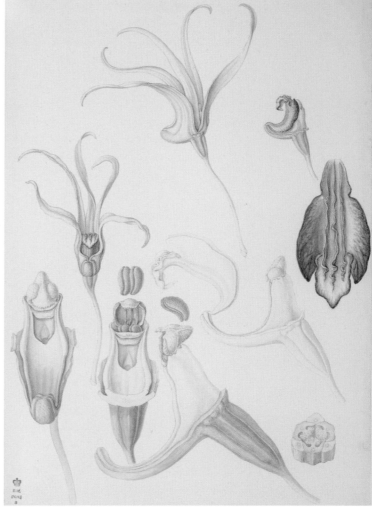

ABOVE: The leaves of the Australian tongue orchid, *Dendrobium linguiforme*, are egg-shaped, thick and fleshy; their grooved surface and overall shape give this orchid both its common and scientific name. The thick leaves help the plant conserve water during dry periods in its sclerophyll forest habitat.

ABOVE: The Australian tongue orchid's spidery white flowers are produced in groups, or inflorescences, from the join of the leaf with the stem. The sepals are long and delicate and give the flowers their overall appearance; the pale purple and yellow petals are smaller and less striking.

OPPOSITE: The flowers of most orchids are upside down; the pedicel or ovary twists so that the uppermost petal is lowermost in flower. These flowers are known as resupinate. The clamshell or cockleshell orchid, *Prosthechea cochleata*, is unusual in being non-resupinate, with the dark, hooded lower petal, called the lip or labellum, remaining as the uppermost part of the flower, arched over the spidery green sepals.

× 2.

Plate 23.

Fitch del.

Lælia superbiens.

Reeve imp.

ABOVE: In his *The Orchidaceae of Mexico and Guatemala*, James Bateman called this showy orchid, *Laelia superbiens*, from Mexico and northern Central America 'the Gorgeous Laelia'. And gorgeous it is, with inflorescences described by its collector Mr Skinner as 'flower stems four yards in length, and bearing upwards of twenty flowers'. Behind the flowers here you can see the large pseudobulbs with their yearly rings of growth, clearly showing their stem origins.

OPPOSITE: Members of the genus *Porroglossum*, like this *Porroglossum muscosum*, are tiny orchids growing in the cloud forests of northern South America. The genus name refers to the hinged labellum or lip (*porro* – far away; *glossum* – tongue), which snaps shut when touched, perhaps enclosing an insect in the flower for a time. But the closure is not permanent. After a time the lip opens again, ready for the next visitor.

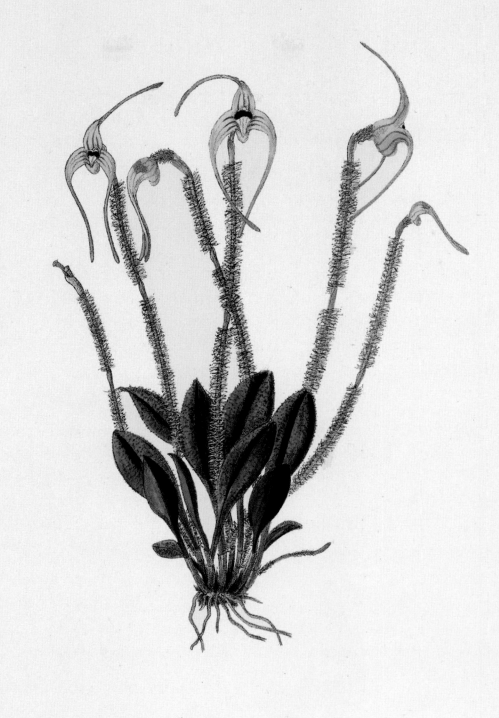
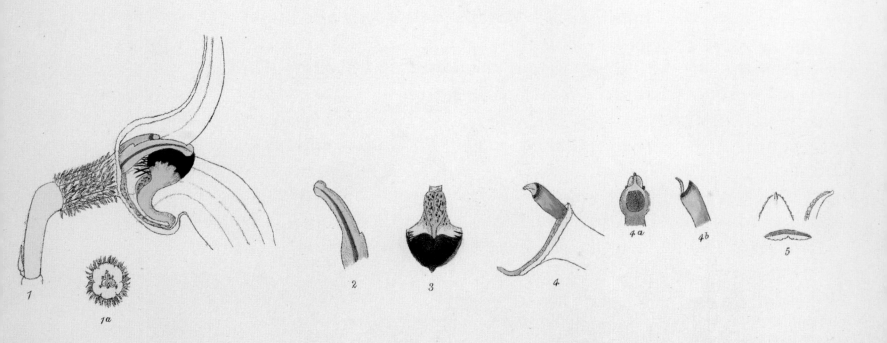

W. Fitch, del.et.lith.

Vincent Brooks, Imp.

ABOVE: The leaves of this orchid, *Luisia psyche*, are not at all leaf-like. Their pointed shape and cylindrical form make them seem more like green awls emerging from the stem. Growing from northeastern India through southeastern Asia, *Luisia psyche* can be found both on trees as an epiphyte and on rocks – orchids growing on rocks are called lithophytic or rupicolous.

Stelis ophioglossoides.
June 15th 1825.

ABOVE: If you don't look closely, the flowers of this tiny orchid, *Stelis ophioglossioides*, look like mere triangles. But focus in, as the artist Franz Bauer did in June 1825, and you can see the tiny petals held inside the triangular outline of the sepals – an orchid in super-miniature. This species was first discovered in the Caribbean by the Austrian explorer Nikolaus von Jacquin and is the type species of *Stelis* whose genus name refers to a mistletoe and the early, but mistaken, idea that orchids were parasites.

ABOVE: The Reverend William Deans Cowan kept careful accurate notes on the habits of all the orchids he painted in Madagascar. But the pictures alone were not enough to describe the new species, like these *Angraecum* species he found. He also collected specimens, these were used in conjunction with his notes and paintings by the British botanist Henry Nicholas Ridley to describe many new species. These specimens are now held in the herbarium at the Natural History Museum for future scientists to examine and compare to new finds.

OPPOSITE: 'While detained at Isabal [Izabal, Guatemala] by the cholera, I quietly took a canoe, and amused myself by a cruise of a few leagues ... in search of our favourite Orchidaceae. I returned home, drenched to the skin, but happy, nevertheless, in the highest degree, for I had discovered a most beautiful plant, and one which I am certain is new to you all.' George Skinner's find was indeed new and was sent to Bateman, who named it *Epidendrum stamfordianum*, for the Earl of Stamford and Warrington, an aristocratic orchid fancier.

Pl. II.

Mrs. Withers del.

EPIDENDRUM STAMFORDIANUM.

M. Gauci, lith.

Pub.d by J. Ridgway & Sons, 169, Piccadilly, Sept.r 1.st 1838.

Printed by C. Hullmandel.

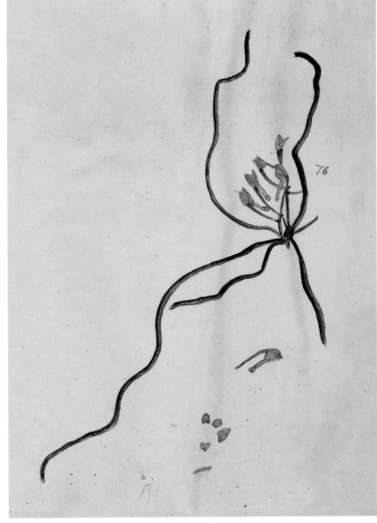

ABOVE: This species of fairy orchid, *Oberonia disticha*, occurs across Africa to Madagascar. Its species name comes from the distichous leaves that are bilaterally compressed and arranged alternately in rows, like iris leaves, but in miniature. An entire plant can be only about a couple of centimetres long. According to William Deans Cowan's careful notes, it grew 'on trees in Ankafana'.

ABOVE: This tiny leafless orchid, *Microcoelia gilpiniae*, endemic to Madagascar, was named for Helen Gilpin, a British missionary and schoolteacher from Bristol in England. She was based in Madagascar from 1869 to 1895 and assisted Mrs Sarah Street at the Girl's School at Faravohitra in Antananarivo, later founding the Home for Girls in Antananarivo. The ending *-iae* on the species name tells us the name honours a woman.

OPPOSITE: The bulb-like protuberances at the base of the leaves, seen here in *Promenaea xanthina*, are called pseudobulbs. They are not bulbs at all – those are structures made of overlapping leaf bases like onions or tulips – but are instead thickened portions of the orchid stem or rhizome. Pseudobulbs can act as storage organs during times of water shortage when leaves dry up.

PROMENÆA CITRINA.

Pl.181.

W.Fitch, del. et lith.

Vincent Brooks, Imp.

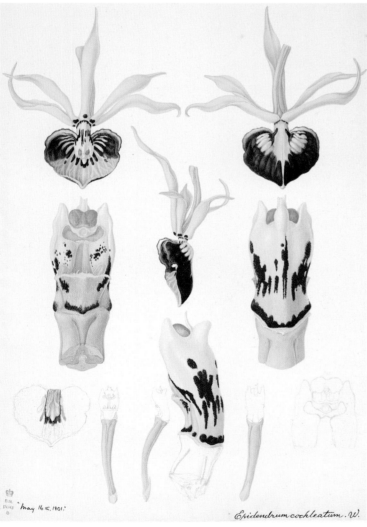

OPPOSITE: Not all epiphytic orchids are delicate balancers on branches. Some are massive indeed – plants of *Grammatophyllum speciosum* have been measured at almost 13 metres (43 feet) in diameter. A plant of this species of tiger orchid that was exhibited at the Crystal Palace in London in 1851 weighed two tonnes! This was a plant for the rich grower with a very large greenhouse in Victorian times, and a real rarity.

LEFT: The green roots of this South Seas epiphyte, *Taeniophyllum fasciola*, are tightly appressed to the branches upon which they grow, looking like a writhing mass of green snakes or worms. This species of *Taeniophyllum* has no leaves and relies on chloroplasts in the roots for sugar production.

LEFT: The clamshell or cockleshell orchid, *Prosthechea cochleata* – so-called for its strikingly shaped upward-pointing labellum – was among the first tropical orchids to flower in Europe. It was named as *Epidendrum cochleatum* by Linnaeus in 1733 from specimens brought from the Caribbean. In Belize, this is known as the black orchid, and is the national flower of the country; it grows all around the Caribbean, including in southern Florida. Franz Bauer depicted these flowers the wrong way up; in nature the purple labellum is uppermost in the flower.

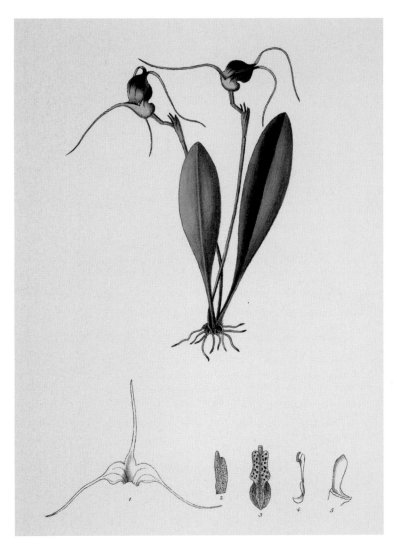

ABOVE: The common name in English for this little epiphyte, *Masdevallia infracta*, is the crooked masdevallia, in reference to the kink in the floral tube. It was first described from plants collected in the wet forests of the Organ Mountains that frame the city of Rio de Janeiro, Brazil, but is also known from Bolivia – a large disjunction for such a tiny plant.

ABOVE: This popular cultivated orchid illustrates one of the growth forms of epiphytic orchids. The stems of the noble dendrobium are not clustered but elongated and grow to 60 centimetres (24 inches) long; flowers are borne at every leaf node. This Himalayan species, *Dendrobium nobile*, has been used extensively in horticulture and is the parent for many hybrid orchids that are among the most popular in the trade.

OPPOSITE: Some species of orchids, such as this *Myrmecophila tibicinis*, are always found growing in ant gardens. Colonies of ants live inside the large, hollow pseudobulbs which are filled with organic debris brought by the foraging ants. The orchids benefit from the increased mineral supply brought to the nest by the ants, but ants also protect the plants from herbivores. Ants that build these aerial nests are usually aggressive and their bites and stings can deter even determined plant collectors. While plant collecting in Panama I pulled down an ant garden to collect the plants in it and the inhabitants rained down into my shirt; I rued my actions for days afterwards!

PL. 30.

M.ᵗˢ Drake, delᵗ

M. Gauci, lith

SCHOMBURGKIA TIBICINIS.

Pub.ᵈ by J. Ridgway & Sons, 169 Piccadilly, Dec.ʳ 1841.

Printed by P. Simon

Roots of love

'These kinds of Dogs stones be of temperature hot and moist, but the greater or fuller stones seemes to have much superfluous windinesse, and therefore being drunk it stirreth up fleshly lust.'

John Gerard, *The Herball,* 1597

CHAP. 110.   *Of Dogs Stones.*

¶ *The Kindes.*

STones or Testicles, as *Dioscorides* saith, are of two sorts, one named *Cynosorchis*, or Dogs Stones; the other, *Orchis Serapias*, or Serapias stones. But because there be many and sundry other sorts differing one from another, I see not how they may be contained vnder these two kinds only: therfore I haue thought good to diuide them as followeth ; the first kind I haue named *Cynosorchis*, or Dogs stones : the second, *Testiculus Morionis*, or Fools stones : the third, *Tragorchis*, or goats stones : the fourth, *Orchis Serapias*, or Serapias stones : the fift, *Testiculus odoratus*, or sweet-smelling Stones, or after *Cordus*, *Testiculus Pumilio*, or Dwarfe stones.

† 1 *Cynosorchis major.*　　　　　　　　† 2 *Cynosorchis major altera.*
Great Dogs stones.　　　　　　　　　　　White Dogs stones.

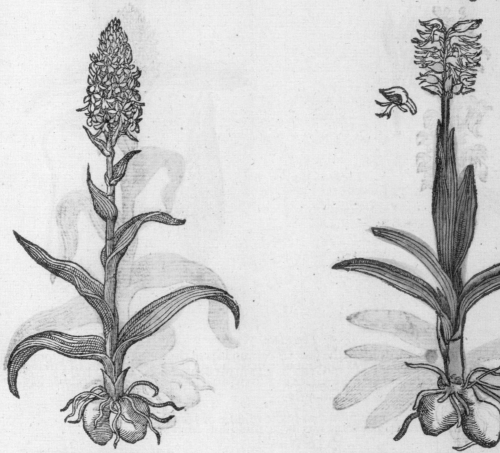

¶ *The Description.*

1 GReat Dogs stones hath foure and sometimes fiue great broad thick leaues, somewhat like those of the garden lilly, but smaller. The stalk riseth vp a foot or more in height ; at the top whereof groweth a thick tuft of carnation or horse-flesh coloured floures, thick and close thrust together, made of many small floures spotted with purple spots, in shape like to an open hood or helmet. And from the hollow place there hangeth forth a certain ragged chiue or tassel, in shape like to the skin of a dog or some such other fourfooted beast. The roots be round like vnto the stones of a dog, or two oliues, one hanging somewhat shorter than the other, whereof the highest or vppermost is the smaller, but fuller and harder. The lowest is the greatest, lightest, and most wrinkled or shriueled, not good for any thing.

2 Whitish Dogs stones hath likewise smooth long broad leaues, but lesser and narrower than those of the first kinde. The stalk is a span long, set with fiue or six leaues clasping or embracing the same round about. His spiky floure is short, thicke, bushy, compact of many small whitish

purple

BOTH THE COMMON and scientific names of orchids (they are in the plant family Orchidaceae) come from the Greek word όρχις (*orchis*), meaning 'testicle'. Hardly the image one associates with an orchid flower. But this refers to the roots of many species of European terrestrial orchids, the underground parts of which in the spring and summer consist of two egg-shaped tubers (sometimes called 'bulbs', but different from bulbs in the botanical sense). Much of the knowledge that Europeans had of plants came from the early Greek and Roman physician-philosophers such as Theophrastus, Pliny and Dioscorides, who were working and writing during the early first century of the common era. In those times, all medicine was derived from plants, so a knowledge of botany was essential, and the early works on plants were largely centred on those used in the treatment of disease and other human conditions.

Theophrastus came from the Greek island of Lesbos and was a student in the Platonic school of philosophy; he took over the management of the Lyceum in Athens when Aristotle was banned. His two works on plants constitute the first real compendium of European knowledge of plant diversity and of the uses of plants by people. They also contain much information on plants of Asia, brought back by those who had fought with Alexander the Great. Among the many plants Theophrastus described is '...at least one plant whose root is said to show both powers [of provoking lust]. This is the so-called salep, which has a double bulb, one large and one small. The larger, given in the milk of a mountain goat, produces more vigour in sexual intercourse; the smaller inhibits and forestalls.' This plant, the 'so-called salep', is an orchid. Just which orchid Theophrastus was describing is not clear; he did travel extensively through Greece, and it could have been a native species there or from further afield. That the same plant could provoke opposite reactions – increasing sexual powers or decreasing them – was not surprising to him, nor is it uncommon in traditional medicine in many cultures. I remember a plant in the Amazon where one dosage caused abortion and a lower dose allowed a woman to conceive. But back to ancient Greece and Rome...

The ideas about the aphrodisiac and anti-aphrodisiac qualities of orchids were copied from Theophrastus by Pliny the Elder, who wrote his *Natural History* in the first century of the common era. Here the tubers are again said to be two, with one harder

OPPOSITE: The dog's stones of early herbals were a mixture of several species of European terrestrial orchids – all ascribed powers to provoke lust.

and larger being 'provocative of lust' and the smaller, softer one being an anti-aphrodisiac. This is how rumours begin... Pliny also describes another plant, again probably an orchid, in which the two tubers stimulate the conception of children of different sexes – the larger one promotes male offspring and the smaller female offspring. This plant was called 'Satyrion' – a name given to any aphrodisiac, but given the description of the two tubers, most likely an orchid. These two properties of orchids became fixed in the materia medica of Europe, largely through the copying and re-copying of the work of Dioscorides.

Pedianus Dioscorides was a doctor of Greek origin serving in the Roman army who lived and wrote in the first century. His book *De Materia Medica* detailed the properties of plants used by people in his day, and throughout the next several centuries was copied and commented upon by generations of herbalists and physicians throughout Europe and beyond. It is not known if he read Theophrastus or Pliny, but his account of the uses of orchids is remarkably similar. He says: 'The root is bulbous, somewhat long, narrow like the olive, double, one part above, the other beneath, one full but the other soft and full of wrinkles. The root is eaten (boiled) like bulbus. It is said that if the bigger root is eaten by men, it makes their offspring males, and the lesser eaten by women makes them conceive females. It is further related that women in Thessalia [the region of Thessaly in central Greece, the Aeolia of Homer's *Odyssey*] give it to drink with goat's milk. The tenderer root is given to encourage venereal diseases, and the dry root to suppress and dissolve venereal diseases.' Pretty similar story, although why one would want to encourage venereal disease is a bit mysterious; maybe he meant it as a warning to be careful not to overdo the use of lust-provoking plants!

As herbals began to be written in local languages, the myths about the properties of orchid tubers grew. John Gerard, whose *Herball* (1597) was one of the first herbals to be published in the English language, stated that the tubers of 'dog's stones' (various species of *Orchis*) were best for stirring up lust, but that 'goat's stones' (*Himantoglossum*) were not used because of 'the stinking and loathsome smell and savour they are possessed with.' The common names of orchids reflected their mythical uses: 'foole's stones' (testicles), 'Knabenkraut' (lad's weed). By the sixteenth century, the use and reuse of Dioscorides had cemented the properties of orchids firmly. The physical

resemblance of orchid tubers to male animal genitals was taken into the new doctrine of signatures, a Christianising philosophy that suggested that the outward appearance of plants indicated how God intended them to be used. In a way, this allowed old myths and uses to continue under new guises, including that of the orchid tuber. Invoking the doctrine of signatures, Oswald Croll in his 1609 book *Basilica Chymical* stated that, 'all species of Orchis, from their similitude of the Testicles, are exciters of the Venerial Faculty, where it is defective; one is dissolved in the liquor of another, the Superior is greater, and fuller, and is powerful in provoking Copulation; the inferior is softer, and withered, inhibiting the Procreative Faculty. Nature industrious in the Generation of Mankind, by this representation signifies, that these are powerful in Venerial Vertues, Conception and Off-spring.' By the early part of the seventeenth century, however, hearsay had begun to go out of fashion. In his 1629 book *Paradisi in Sole Paradisus Terrestris*, English herbalist John Parkinson repeated Dioscorides, but with the caveat that no one he knew had actually experimented with the method to find out if it worked. So much for the Viagra of the late Middle Ages. By the early 1900s, herbals were touting the use of fresh tubers to promote true love, and withered ones to check wrong passions, a slightly less racy use for these underground parts.

Thus, the myths were pretty much debunked, but what of the tubers of these terrestrial orchids themselves? The observation that one was firmer and harder and other softer and withered is very accurate; in fact, these underground organs are storage

ABOVE: *Himantoglossum hircinum*. 'The floures which grow in this bulb or tuft be very small, in form alike unto a Lizard, because of the twisted or writhen tails, and spotted heads. Every of the small flooures is at the first is like a round close huske, of the bignesse of a pease, which when it openeth there commeth out of it a little long and tender spurre or taile, white towards the setting of it to the floure, the rest spotted with red dashes, having upon each side a small thing adjoining unto it like toa a little let or foot, the rest of the said taile is twisted crookedly about, and hangeth downward. The whole plant hath a rank or stinking smel or savour like the smell of a Goat, whereof it took its name.' Gerard, *The Herball*, 1597.

organs for the plant. In those species with two tubers, one is produced each year. The withered one is last year's that has supported the current year's growth, while the firmer, plumper one is full of starchy material laid down for next year. Both tubers are full of a starchy polysaccharide called glucomannan that serves as a storage compound for the plant. It is used by people nowadays, not for various love or virility potions, but as the basis for a nutritious drink called salep, although in some places orchid tubers are still considered an aphrodisiac. Salep has a long history; Theophrastus mentioned it as the tuber itself σαλέπι ('salepi'), and the word itself has Arabic origins – transliterated as sahleb, and eventually as saloop in the coffee houses of London of the seventeenth and eighteenth centuries, where it was popular before coffee became the drink of choice.

Salep is made from the ground dried tubers of a variety of species of terrestrial orchids, originally from the genus *Orchis*, but a wide variety of tuber-bearing species are also used. The word itself refers to both the ground powder and to the drink itself. Salep is used commercially in ice cream as a thickening agent and in candy, and it is still a popular drink today in southeastern Europe, mostly Greece and Turkey. The tubers are harvested from the wild and then dried and ground, and the resulting powder is added to milk with sugar and various flavourings. The withered tubers are not preferred for salep production, as they contain little starch, and they are usually discarded. Salep powder is used locally as a warming drink during the cold winter

ABOVE: *Cypripedium calceolus*. The slipper orchids are among the most prized by collectors, and overcollection in the wild has driven the lady's slipper in Britain almost to extinction. Use of genetic testing has helped to understand the origins of populations in both Britain and Europe, and is used to ensure genetic bottlenecks are minimized during introduction programmes that have today successfully established populations of this species. Even though success has been achieved, some locations are kept secret to deter decimation by unscrupulous collectors.

months, and as treatment for gastrointestinal disorders, especially in infants. It was used in northern Greece for porridge and has been considered a traditional staple. A renewed interest in traditional remedies and local traditions has meant that demand for high-quality salep is on the increase.

This increased interest is a problem – not a medical one, but one of sustainability. A kilogram of salep powder requires some three to four thousand tubers, thus three to four thousand individual plants – each harvested from the wild. Estimates of millions of orchids harvested annually mean that over-harvesting is common, and the availability of the product is at issue. In addition, all orchids are protected by law, both nationally and internationally. The Convention on International Trade in Endangered Species of Wild Fauna and Flora (CITES) lists all members of the orchid family as protected under Annex I – no trade permitted. But local use is culturally embedded, and the harvesting of salep is an important source of income for local people. A conservation conundrum.

The underground parts of terrestrial orchids are not only used in Europe but are also an important component in Traditional Chinese Medicine (TCM). Several species of orchid are used, but the tubers of *Gastrodia elata* – a ground orchid that grows in forests in northern China – are favoured. The tubers are known as *tiān má* and are a 'top grade' (*shàng pǐn*) medicine: high quality and non-toxic. The tubers are used in the treatment of depression, cognitive disorders and a wide variety of related problems. TCM is the primary source of health care for most people in China, and the properties of medicinal herbs have been well studied and their efficacy shown. The industry is highly regulated, and issues related to over-harvesting are coming to the fore, sparking efforts to cultivate key top-grade remedies. *Gastrodia* has a complex life cycle involving specialized wood-rotting fungi (see p. 106), but cultivation has begun for use in TCM. However, as with the European orchids used in salep, wild-harvested plants are often considered higher quality, thus driving over-harvesting through demand for what is regarded as the highest quality product. This is a vicious circle.

Since orchid tubers are pretty amorphous and non-descript when dried, and even more so when ground, the very composition of products on sale can be altered in many ways. Scientists have used a technique called DNA meta-barcoding to test the

 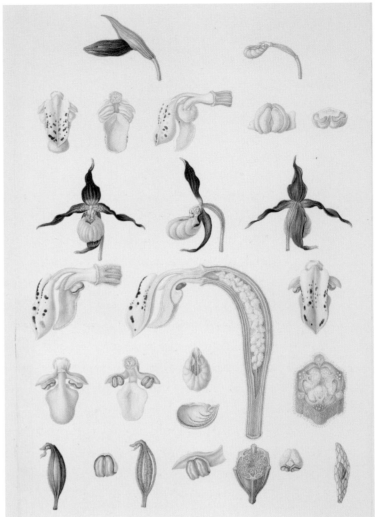

composition of orchid-derived products. Advances in technology allow assays to be made of both individual objects and complex mixtures to ascertain if they are what they are said to be. Often they are not. Tubers sold in the relatively unregulated London Chinese medicine markets have been found to be made from carved potatoes, and salep tested in various parts of southeastern Europe and Asia contained multiple species of orchid. Some was also completely free of orchid DNA. Salep powder in markets is often adulterated with synthetic polysaccharides or with other plant-derived starches. DNA barcoding showed that most of the salep-containing processed foods contained no orchid residues at all; so, many of the products are not all what they seem. Nonetheless, pressure on wild orchid populations continues, as demand increases for traditional, organic products in both food and medicine.

Medicinal plants are not the only ones at risk from over-harvesting. Orchids have long fascinated collectors, and some species of terrestrial orchids have been collected nearly to extinction for use in gardens by enthusiasts. The European slipper orchid,

ABOVE LEFT: The rare ram's head slipper orchid, *Cypripedium arietinum*, grows in swampy alkaline soil around the Great Lakes in the United States and Canada.

ABOVE RIGHT: The common name slipper orchid derives from the sack-like labellum, or lip, of these flowers.

Cypripedium calceolus, was once widespread across northern Britain, but by the late-twentieth century wild populations were reduced to a single individual by over-collection, both for herbarium specimens and cultivation in private gardens. A gardener and nature enthusiast called Robert Gathorne-Hardy poetically pointed the finger: 'this, the most spectacular of British flowers, is the one certainly attested victim of those much abused people, the collectors. . . . . but it was garden greed – that must be allowed – which swept away the lady's slipper.'

Work by scientists at the Royal Botanic Gardens, Kew, pioneered seed germination and propagation from seed of this species, and a programme of reintroduction has resulted in its re-establishment in several locations, many of which are kept secret to prevent poaching. And poaching is the right word – just like poaching of animal wildlife, the illegal and unregulated removal of wild plants from their environments is a risk to biodiversity and to the survival of these species as integral parts of Earth's ecosystems. The illegal plant trade has received far less attention than trade in animals such as pangolins, rhinos and elephants, but it is every bit as widespread, and every bit as lucrative for those who are involved. Fortunately, both scientists and conservation organisations are beginning to see plants as wildlife too, and global efforts to combat the illegal trade in wild-harvested plants are on the increase.

What about the traditional use of plants by local people? And the over-harvesting not only for trade, but for local use? Here perhaps horticulture can provide us with some solutions to the loss of populations and species collected in the wild. The focus on endangered species means that methods of propagation from seed and by forms of vegetative (asexual) propagation – those of the former were developed for the lady's slipper orchids in Britain – can be developed for the (relatively) mass production of orchids for use by people. Such methods will also allow us to perhaps reintroduce populations to the wild where they once existed, but only if we can reduce demand for wild-collected plants in the first place. Wild orchids epitomise the conundrum of conservation in today's world – just fencing off areas will not in itself conserve biodiversity; people are an integral part of the equation. Changing the world involves changing ourselves.

Plate XCIV.

Reeve Benham & Reeve, imp.

Habenaria gigantea.

ABOVE: Not all terrestrial orchids are from the temperate zone, nor are they all small seasonal plants. The flowers of *Sobralia* species, such as this *Sobralia macrantha*, are large and showy and the plants have the habit of bamboo, often growing to several metres tall on rocky slopes and open areas along roads in the New World tropics.

OPPOSITE: Georg Everhard Rumphius, the blind seer of Ambon, named a close relative of this spectacular orchid 'Flos susannae' after his wife, Susanna, a local woman 'who during her life was my first spouse and helpmate in the finding of herbs and plants, and because she discovered it'. Rumphius lived in what is now Indonesia in the seventeenth century; he became blind in later life and completed his botanical works by dictation and therefore from memory. His descriptions of tropical Asian plants were among the first to reach Europe.

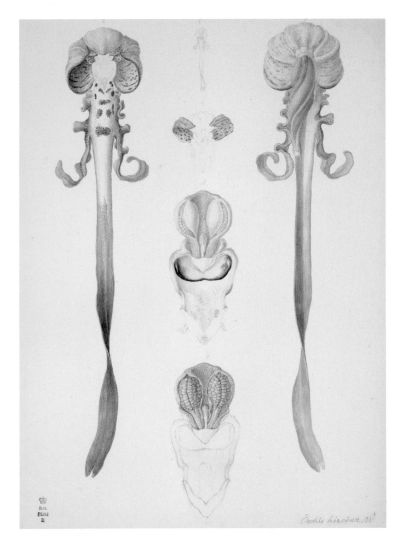

ABOVE: This plant, *Himantoglossum hircinum*, is today known as the lizard orchid in English, while the common names in Germany (Bocks Riemenzunge) and France (l'orchis du bouc) reflect its goaty smell. John Gerard had very carefully observed the flowers, but had trouble describing the unusual parts (see caption page 43).

ABOVE: This pale-flowered orchid, *Orchis pallens*, grows in the meadows of the high Alps. One of the twin tubers persists over the winter, providing nutrition for plant growth in the spring before the leaves are fully formed and ready to photosynthesize. This one shrinks over the course of the summer, while the other one is full of storage tissue laid down during the growing season, thus allowing the plant to persist from year to year.

OPPOSITE: *Neotinea ustulata*. European orchids are among the most studied plants in the world. This can be good when evolutionary relationships are clarified, although name changes can sometimes upset people. But this popularity can have a more problematic side too. Rampant over-splitting – recognition of many microspecies on very tiny characters – can obscure real patterns of diversity and impede conservation of plants and habitats that are really at risk.

PLATE CLXIV

Tawny Thrush.

TURDUS WILSONII. *Male. Habenaria Lacera – Cornus Canadensis.*

Drawn from Nature by J.J.Audubon, F.R.S. F.L.S.

Engraved, Printed & Coloured by R.Havell. London. 1833.

OPPOSITE: John James Audubon's *The Birds of America* is a masterpiece of natural history publication. The birds were drawn not from life, but in life-like poses held up with wires and sticks, but they were placed in realistic vegetation. This plate depicts the veery or tawny thrush, that Audubon drew in Maine, along with several plants that occur in that region as well, such as the dwarf cornel, *Cornus canadensis*, and the fringed orchid, *Platanthera leucophaea*.

LEFT: The grass-like leaves of this terrestrial orchid, *Orthoceras strictum*, from New Zealand, found by Joseph Banks and Daniel Solander on Captain Cook's epic voyage on HMS *Endeavour*, show the orchid family's relationships to other monocotyledonous plants like grasses and sedges; once in flower though there is no mistaking this plant is an orchid!

ABOVE: The bottle-gourd Indian crocus, *Pleione × lagenaria*, is not a crocus at all, but a terrestrial or lithophytic orchid from the Himalaya in northeastern India and southern China. It is a naturally occurring hybrid but is easier to grow in cultivation than either of its parents. The strange gourd-shaped leaves are actually stems (pseudobulbs) and persist for only a year; the plant flowers when the true leaves have dropped.

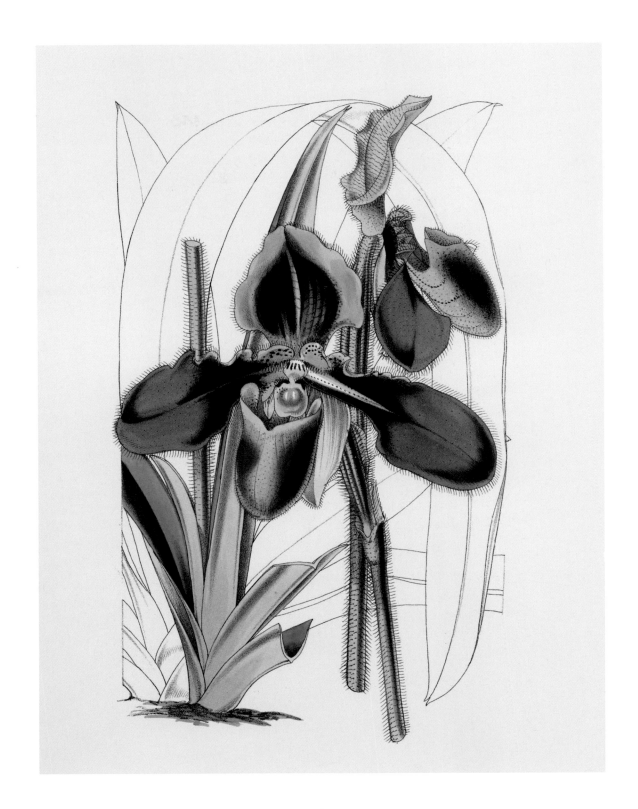

ABOVE: This Chinese slipper orchid, *Paphiopedilum hirsutissimum*, is said to be one of the easiest to grow inside without special greenhouse treatment; its softly hairy flower parts are striking and unusual and it flowers year-round. When originally described it was thought to come from Java, but its real native range is from Assam in India to southern China.

Nugent Fitch del. et lith.

CYPRIPEDIUM STONEI.

J.N.Fitch imp.

OPPOSITE: *Paphiopedilum stonei* takes its species name from Robert Stone, who was the gardener for John Day – the son of a wealthy wine merchant who became one of the great Victorian orchid fanciers. With his wealth he bought orchids from nurserymen, but also from collectors, and grew them in his high-tech, for the time, heated orchid house at High Cross, Tottenham, London. It was surely Stone who brought the plant into flower so William Hooker could illustrate and describe it.

LEFT: The white moccasin flower or small white lady's slipper, *Cypripedium candidum*, takes its common name from the inflated white pouch formed by the flower's labellum or lip, which greatly contrasts with the dark brown or reddish-brown sepals. This North American species grows in prairies and bogs, and sometimes even along railway lines in the open sun.

LEFT: Of the 130 orchid species native to the tiny island of Hong Kong, ten have been prioritized for global conservation efforts and are being propagated from seed for reintroduction to the wild. One of these is *Paphiopedilum purpuratum*. Only known from a handful of plants, it is severely threatened by overcollection – it is a popular parent for commercial hybrids of slipper orchids. Seeds are grown on artificial media and can take up to two years before they are strong enough to be planted out onto soil or rocks.

ABOVE: This terrestrial orchid, *Cynorkis lilacina*, was described by Henry Nicholas Ridley based on specimens collected by the Scottish missionary William Deans Cowan who lived in Madagascar associated with the London Missionary Society from 1874 to 1881. These watercolours accompany the specimens and give crucial details about flower colour, habit and smell – features that are lost in dried specimens.

ABOVE: This Malagasy orchid, *Cynorkis bimaculata*, has been assessed as critically endangered by the International Union for Conservation of Nature (IUCN); it is known from a single location. Terrestrial orchid habitats are threatened by habitat destruction in the tropics everywhere – I wonder what Henry Ridley, who first described this species, would think of his enthusiasm for oil palm plantations now. Oil palm cultivation has brought destruction to forests in Asia and its spread to Africa does not bode well for forest organisms there.

OPPOSITE: This cane orchid, *Thunia bensoniae*, bears large flowers at the end of bamboo-like stalks. It is deciduous and loses its leaves every year; new shoots grow from the base, each of which has a few flowers at its tip. The species honours the wife of Colonel Robson Benson, who was the British resident to the court of Burma in the late 1800s and had sent plants to Joseph Dalton Hooker at Kew who said, 'At Colonel Benson's request it is named after his lady, and few more beautiful plants have ever borne a lady's name.'

PL.67

J.Nugent Fitch del.et lith.

THUNIA BENSONIÆ.

J.Nugent Fitch imp.

Pl. 170.

W. Fitch, del. et lith.

Vincent Brooks, Imp.

ABOVE: John Lindley was the Victorian era's premier orchidologist – quite something during the flourishing of orchid-mania. He did not confine himself to the big showy epiphytes, but also described more delicate species like this *Cynorkis uniflora*. In the 1840s Lindley was responsible for forcing a discussion of the fate of the Royal Botanic Gardens at Kew by publicizing its imminent dismantling by the Government – the outcry saved the gardens from destruction.

OPPOSITE: The leafy orchid, *Dactylorhiza foliosa*, is endemic to the island of Madeira – it grows nowhere else in the world, at least naturally. Many plant species are known only from Macaronesia – as the Atlantic islands of Madeira, the Azores and the Canaries are known – and they were among the first non-European plants to be seen by European explorers setting off to explore the world.

ABOVE: Orchids grow everywhere – this species, *Codonorchis lessonii*, is from the far southern lands of Tierra del Fuego in Patagonia, a harsh and unforgiving environment for plants and people alike. During the passage of the area by HMS *Beagle*, Charles Darwin remarked that: 'The climate is certainly wretched: the summer solstice was now passed, yet every day snow fell on the hills, and in the valleys there was rain, accompanied by sleet. From the damp and boisterous state of the atmosphere, not cheered by a gleam of sunshine, one fancied the climate even worse than it really was.'

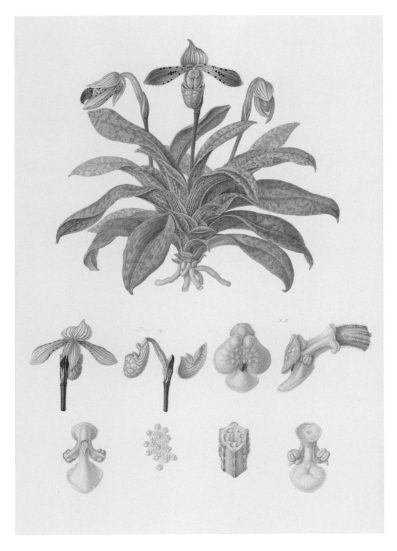

ABOVE: This orchid, *Paphiopedilum venustum*, was called the comely ladies slipper when it was formally described; its species name refers to the Roman goddess Venus. It came into cultivation in Europe from plants sent from the Calcutta Botanic Garden, whose Director, Nathaniel Wallich, mentioned them as a third new species of slipper orchid called 'Cypripedium venustum' in a letter to another celebrated explorer of the Himalayan forests, Francis Buchanan-Hamilton. The describer of this species took up Wallich's apt name. Its native range is in northeastern India, Nepal and Bhutan, in 'some of the wettest places on earth'.

ABOVE: The genus name for this European orchid, *Dactlyorhiza maculata*, means finger (dactyl-) roots (-rhiza) and refers to the characteristic shape of the tubers – like the fingers of a hand, rather than two round or oval balls as in many other European terrestrial orchids.

OPPOSITE: The South Sea islands are home to many epiphytic orchids, but terrestrial orchids also occur. This inverted boot orchid, *Crepidium resupinatum*, with its thin membranous leaves, so different to those of the epiphytes, was sketched in Tahiti by Sydney Parkinson, the artist who accompanied Joseph Banks and Daniel Solander on the voyage of HMS *Endeavour*, led by Captain James Cook of the British Royal Navy.

Epipactis purpurea.

Otaheite

Sydney Parkinson pinx 1769

ABOVE: The peinktrewwa or geelkappie is a plant of the winter-rainfall habitats of fynbos and renosterveld in the diverse Cape region of South Africa, usually growing on dry sandstone or clay flats. The genus name comes from σάτυρος (satyros) the two-horned satyr of Greek mythology with the body and face of a man and the tail and ears of a horse or goat; satyrs were notoriously bawdy and obscene. The labellum of *Satryium erectum* is hooded and held uppermost, with its two spurs it somewhat fancifully resembles a goat's head or a devil with a pair of horns.

OPPOSITE: With their single tiny leaf and dark coloured, short-lived flowers, helmet orchids, *Corybas fimbriatus*, are seldom collected or even noticed as they grow on the forest floor. Their new (sometimes called replacement) tubers are formed at the end of an elongated 'dropper' allowing plants from seasonal temperate zone habitats of eastern Australia to survive the dry season well underground, avoiding the extremes of heat and low humidity nearer the surface.

ABOVE: The rattlesnake plantain – the common name of this orchid, *Goodyera pubescens*, in North America – grows by spreading rhizomes and can form huge colonies of veined leaves flat on the forest floor in pine woods with acid soils. Coming upon a group of these plants in the forest is always exciting, even if they are not in bloom.

Monstrous forms

'But will the rarity of the Orchidaceae, or the care and
attention they require, suffice to explain the strange
power of fascination which they are felt to possess? [....]
where, but in the marvellous structure, the grotesque
conformation and imitative nature of their flowers?
Yes; here we have that which is more than sufficient
to explain all the wonder and admiration they
have excited.'

James Bateman,
The Orchidaceae of Mexico and Guatemala, 1843, p.6

ABOVE: Lady Grey of Groby's fanciful imagining of orchids as a menagerie of animals underlines their strange and wonderful forms and appeals directly to the Romantic imagination.

O RCHID FLOWERS FASCINATE. Their sheer exuberance of form and strangeness of shape led great American orchid biologist Oakes Ames, for whom the orchid herbarium at Harvard University is named, to quote late-fifteenth century Dutch merchant Jakob Breyne: 'Nature has formed orchid flowers in such a way that, unless they make us laugh, they excite the greatest admiration.' With their almost sculpted forms, unrecognizable flower parts and long lives – some *Dendrobium* flowers can last up to a month if not pollinated and the flowers of my supermarket *Phalaenopsis* lasted for four months in my kitchen – orchid flowers seem unreal. It is not surprising then that in 1843 James Bateman quoted Milton at the end of his magnum opus *The Orchidaceae of Mexico and Guatemala*:

> '...Nature breeds
> Perverse, all monstrous, all prodigious things
> Abominable, unutterable, and worse
> Than fables yet have feigned, or fear conceived
> Gorgons, hydras and chimeras dire'

These 'monstrous' forms – so entrancing to collectors and artists alike – are the result of evolution by natural selection working through the interactions of plants and their pollinators, even though they do seem quite spectacularly bizarre. When Charles Darwin finished his great book (for him only an abstract and shorter than he would have liked!) *On the Origin of Species by Means of Natural Selection* in 1859, he had laid the underpinnings of a new way of looking at the world. The natural world as we see it was not made according to a plan; instead, plants and animals were constantly locked in a struggle for existence, and those that reproduced better went on to thrive and reproduce again, thus causing change over generations. But Darwin was worried about evidence and 'facts'; for him the evidence of his theory was all around him, but he was a stickler for the marshalling of fact after fact – detailed observations of evolution at work in the natural world. So in the turmoil resulting from the publication of the *Origin* he set to work on organisms that had fascinated him for some time, organisms that were abundant in the woods around his home, Down House, and that he cultivated in his glasshouses: orchids.

Although it is often the voyage of HMS *Beagle* that is said to have opened Darwin's eyes to his explanation for how change occurs in nature, it is really his minutely detailed observations of plants and animals in their natural habitats, day-to-day, that cemented his ideas. Charles Darwin had long had an interest in plant reproduction; one of his first scientific publications treated the pollination of flowers by bumblebees. Darwin's first book post-*Origin* and the first of his six books devoted exclusively to plants has the long-winded title of *On the Various Contrivances by which British and Foreign Orchids are Fertilised by Insects, and on the Good Effects of Intercrossing* (1862). He had begun experimenting with orchid pollination, and in 1860 he submitted a letter to the horticultural magazine *The Gardener's Chronicle and Agricultural Gazette* requesting information on the pollination of British native orchids, about which he had already amassed quite a bit of information:

'I could give many facts showing how effectually insects do their work; two cases will here suffice; in a plant of *Orchis maculata* [now called *Dactylorhiza maculata* or *D. fuchsii*] with 44 flowers open, the 12 upper ones, which were not quite mature, had not one pollen-mass removed, whereas every one of the 32 lower flowers had one or both pollen-masses removed; in a plant of *Gymnadenia conopsea* with 54 open flowers, 52 had their pollen-masses removed. I have repeatedly observed in various Orchids grains of pollen, and in one case *three* whole pollen-masses on the stigmatic surface of a flower, which still retained its own two pollen-masses; and as often, or even oftener, I have found flowers with the pollen-masses removed, but with no pollen on their stigmas. These facts clearly show that each flower is often, or even generally, fertilised by the pollen brought by insects from another flower or plant. I may add that after observing our Orchids during many years, I have never seen a bee or any other diurnal insect (excepting once a butterfly) visit them; therefore I have no doubt that moths are the priests who perform the marriage ceremony.'

And still he wanted more facts. With this letter he hoped to drive a nail into the idea of Robert Brown (first Keeper of Botany at the Natural History Museum in London) that most British orchids were self-pollinated.

Darwin kept at it; his orchid book has page after page of detail about the floral structure of orchids and about how the pollen masses are affixed to insects and moved from plant to plant. He rarely saw this happening in the wild, but he used pencils and pins and sticks to prod and poke flowers to see how orchid pollination worked, and he encouraged his readers to do the same. He not only used the orchids growing in the woods around Down House, but he was also supplied with flowers of exotic tropical species by his friend and Director of the Botanic Gardens at Kew, Joseph Dalton Hooker. Both Hooker and American botanist Asa Gray had been somewhat unconvinced by the arguments in the *Origin*, but the orchid book put paid to that – Gray wrote to congratulate Darwin on his 'beautiful flank movement with the Orchid book' that was gradually winning over the detractors of the theory of evolution. To see what Darwin saw, one must first have a picture of how the orchid flower is constructed. Orchids are monocotyledons – a monophyletic group descended from a common ancestor in which flower parts usually occur in groups of three. Three sepals, three petals, three or multiples of three male organs (stamens) and three female parts (carpels). The carpels are fused into an inferior ovary – not inferior in the sense of being worse but so-called because it is positioned below the other flower parts. But look at an orchid flower and that pattern of threes seems a stretch of the imagination. Most orchid flowers have the male and female organs (anthers, style and stigma) fused in complex ways to make a structure in the centre of the flower called the column, the form of which is highly variable. The petals and sepals are variously fused and shaped to make the incredible diversity of orchid flower forms, but it is the column that really makes most orchid flowers look so peculiar to our eyes. The column holds not only the stigma, or the receptive tissue into

ABOVE: These Southeast Asian orchids, *Neuwiedia veratrifolia*, are members of a lineage that is the first branch on the orchid family tree; they do not have a column, but instead have the anthers and style separate.

which pollen grains grow, but also the anthers, or male parts of the flower. The anthers of most orchids do not contain powdery, loose pollen grains as do most other flowering plants, but instead the pollen is clumped into discrete waxy masses called pollinia. In many orchids the pollinia are attached to a basal sticky structure – the viscidium – by a thin stalk, making the whole structure, called the pollinarium, look like a tiny club. This package is delivered to flower visitors of these species and is usually attached on a very specific part of the pollinator body. The terms Darwin used were different from those used today, but what he observed was that the position of the pollinia on a visiting insect was determined by the attachment of the viscidium, and that the drying of the viscidium made the pollinia change orientation – perfectly placing them for contacting the stigma of the next flower visited.

So far so good. The visiting insects deposit pollen in just the right place to effect fertilization, but what do the insects get out of it? Animal pollinators are often attracted to flowers by the promise of a reward, many times in the form of sweet sugary nectar solutions. Nectar is often hidden away in the flower – at the bottom of long tubular flowers or in elongated spurs. Only a certain method of entering the flower will get the insect what it's after. Many orchid flowers have spurs, tubular extensions of one of the sepals or petals that hold nectar away from both the drying power of the air and from 'thieves' who would collect it without bringing or taking pollen to effect fertilization and thus reproduction.

In January 1862, while just finishing his orchid book, Darwin was sent an extraordinary orchid from Madagascar by James Bateman – *Angraecum sesquipedale*. A magnificent

ABOVE: The orchid column is composed of variously fused floral whorls as in this *Dendrobium polyanthum*; both male and female parts are held in a single structure that is usually quite stiff and solid.

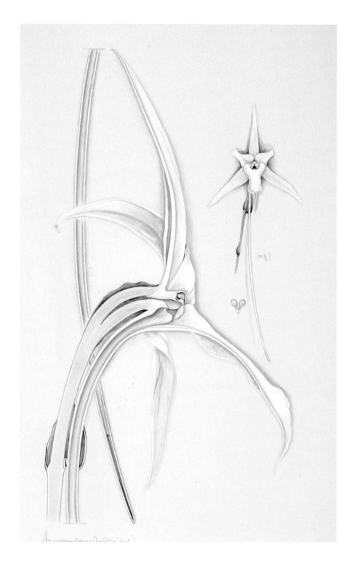

white flower about 10 centimetres (3.9 inches) across 'like stars formed of wax', it had a fabulously long spur that was if not quite 'sesquipedale' (one and a half feet) was long enough to defy the imagination. What insect could reach the nectar held in the bottom part of a tube that was 28 centimetres (11 inches) long? The idea of a moth with a tongue long enough to reach the bottom of this tube was tested by Darwin, who poked and prodded Bateman's gift until he was convinced that only a large-bodied, long-tongued moth could do the trick. But no entomologist had ever seen such a moth. Alfred Russel Wallace supported his mentor and co-discoverer of evolution by natural selection, publishing an article in 1867 in which he stated in a footnote: 'That such a moth exists in Madagascar may be safely predicted; and naturalists who visit that island should search for it with as much confidence as astronomers searched for the planet Neptune,--and they will be equally successful!' It was only 20 years after Darwin's death that the predicted moth was first collected in Madagascar and then named by Walter Rothschild and Karl Jordan – *Xanthopan morganii praedicta* – 'praedicta' for Darwin's prediction. Alfred Russel Wallace was still alive in 1903 to see this prediction come right. It was almost a hundred more years before the moth was observed visiting *Angraecum sesquipedale*, and filmed in the act, thus supporting its putative role in the pollination of this extraordinary flower.

Not all *Angraecum* species are pollinated by long-tongued moths, however; the genus has diversified hugely on the island of Madagascar and in the Mascarene Islands, largely through variation in flower shape based on pollinator difference. Orchids are pollinated by many sorts of insect – bees, moths, butterflies, flies. In 2010, nocturnal cricket pollination was reported for a species of *Angraecum* (*Angraecum*

ABOVE: The nectar spur of *Angraeacum sesquipedale* only has sugary solution in the bottom few centimetres; a long tongue is needed to access this reward.

cadetii) on the island of Réunion. But for careful, Darwin-style observation, one might have assumed that these *Glomeremus* crickets were just cruising from flower to flower, damaging floral structures or robbing nectar from the short spurs. The crickets are probably attracted to the strong scent emitted by the flowers, and do obtain nectar, but at the same time they deposit pollinia in just the right place for fertilization. The case of *Angraecum cadetii* is every bit as wonderful an example of plant-animal interaction driving evolution of form as that of *Angraecum sesquipedale*.

Scent is often associated with night-flowering plants – production of highly volatile chemical compounds attracts insects, and sometimes other animals, to flowers in the absence of other cues such as colour or patterning on petals. Night-flowering orchids such as *Angraecum* are highly scented, but the truly fantastical orchid scents are produced during the day. In the tropics many species of orchids do not produce nectar as a reward for their pollinators; instead, they produce fragrances that are collected by male bees of the family Euglossidae, the 'orchid bees'. Euglossine bees are, in my view, among the most beautiful of bees; often brightly coloured and metallic, they can be large and can fly extremely long distances. The fragrance collected at orchid flowers is used by the bees to create a pheromone – scent – that attracts females. The pheromone of any given bee is a blend of volatile compounds that he collects from many plants via a fatty secretion from glands in his head, which he sprays on the flower to release its fragrances. These chemicals are then collected using special brushes on the forelegs and stored in hollow pouches on the hind legs that act like a sponge to keep the perfumes in place. The payload of these spectacular bees is so highly sought after that it is even removed from the hind legs of dead male bees by other males of the same species.

Visiting orchids to collect fragrance is done only by males – female bees collect nectar from a wide variety of other flowers, including nectar-producing orchids. The fragrances produced by orchids that attract euglossine bees are incredibly complex. One species of *Catasetum* from Panama was recorded with as many as 77 different chemical compounds in its offer. The mechanics of fragrance collection have been characterized as 'drop-through' (*Stanhopea*), 'slide-through' (*Gongora*) and 'scratch and sniff' (*Coryanthes* and *Catasetum*). Stanhopeas are known as the 'upside-down'

OPPOSITE: This Brazilian species, *Stanhopea insignis*, was the first of the spectacular stanhopeas to be cultivated in Britain and was described in 1829. Franz Bauer made this painting in 1827 from those first flowering individuals.

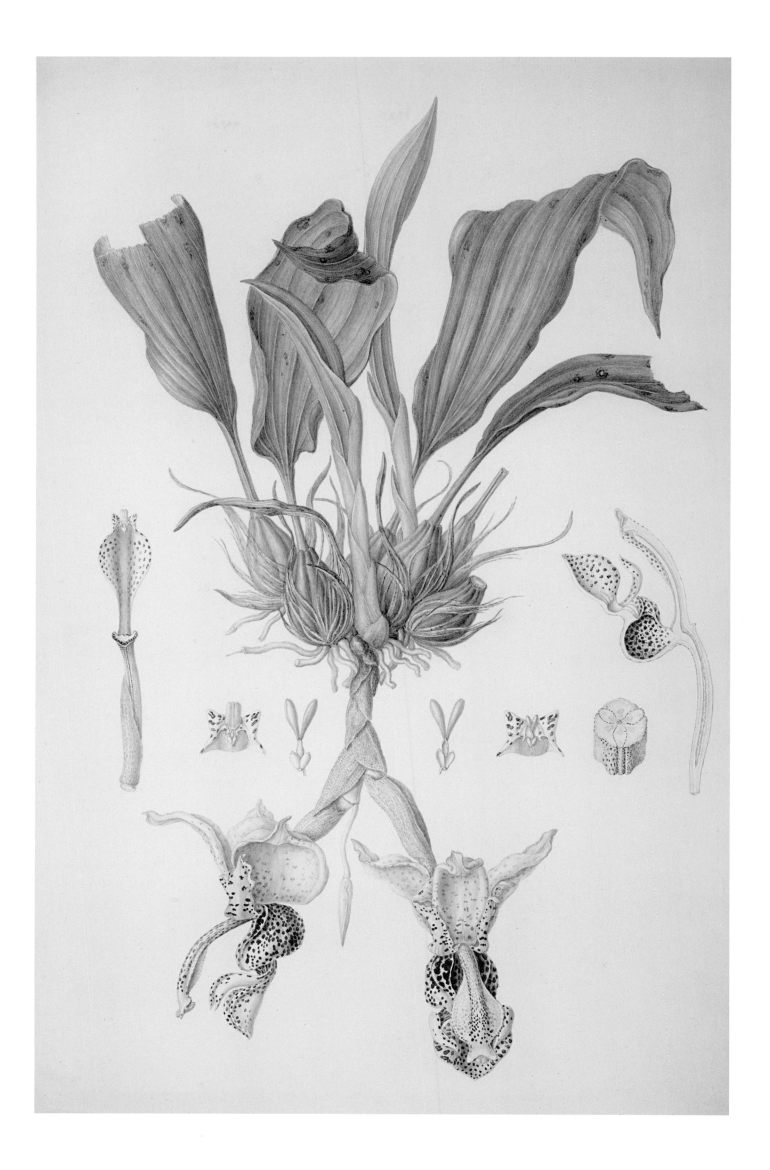

Disa grandiflora

orchids, for their flowers that hang down from the bottom of pots when they are cultivated in greenhouses. Male euglossine bees visit the flowers to collect fragrance that is held deep inside the flower on a structure called the hypochile (the bottom-most part of the lip, or central petal, of the flower). As he collects the fragrance and packs it onto his hind legs, the bee loses his grip on the slippery lip and slides down backwards, touching the column on his way and detaching the pollinia, which are then deposited on his body in a particular place as he ignominiously exits the flower. On visiting the next flower, the same thing happens, but the previous pollinia are deposited on the stigma.

The variations on euglossine bee-orchid interactions are truly testament to the powers of natural selection. But the orchids depend on the bees more than the bees depend on the orchids, so this is really an asymmetrical relationship. David Roubik, with whom I trapped for orchid bees in Panama in the 1980s, says 'orchid bees do not need orchids'. In fact, these bees collect scent from other plants such as aroids and even fungi, and even can be attracted to various scents on pieces of paper: some nice ones such as benzyl acetate (a constituent of jasmine, ylang-ylang and neroli perfumes); other nasty ones such as skatole (smelling of faeces, boar taint or coal tar). Beauty is in the eye of the beholder!

Many orchids offer their floral visitors and pollinators no reward at all – they have lovely flowers that attract insects, but these produce no nectar or any other substance

ABOVE: The red disa, also known as the pride of Table Mountain, is endemic to the sandstone table mountains of South Africa. It offers no rewards to its pollinator, the mountain pride butterfly, *Meneris tulbaghia*, but mimics other flowers in the area that have nectar. These plant guilds are often specialized for a particular pollinator and are a good example of convergent floral evolution among guild members.

that the visitors could collect. This is pollination by deceit; the flower looks like it is a source of food or other reward, but it is not. Some of these species, such as the common spotted orchid, even have spurs like those that in other flowers hold a nectar reward, but in probing the spur with their tongues trying to find the nectar, visitors to these deceptive orchids push against the column and come away with no reward, but with the pollinia firmly attached to their heads. Darwin was puzzled by these food-deceptive orchids, of which there are many in the British Isles, and wondered how such a mechanism was maintained in nature. Why would bees just not learn to avoid them? In the temperate zones these orchids are usually visited in early spring by naïve bees just emerged from the pupa, who soon learn to avoid these non-rewarding plants — although if the orchids are at low density the bees forget quickly! Food-deceivers also flower earlier than do their rewarding counterparts, supporting the idea that they are taking advantage of naïve pollinators and avoiding competition with rewarding plants in the same habitats. But it can work the other way around as well for these deceivers. In South Africa, food-deceptive orchids are members of 'guilds' of similarly coloured flowers that are all pollinated by the same insects. Here orchids are the bit players, as the numbers of other plants in the guild drives the numbers of orchid plants, and the orchids mimic the rewarding plants in the ecosystem. So again, orchids are dependent upon other species that are not dependent upon them.

Pl. 31.

ABOVE: Female flowers of *Catasetum* species, like this *Catasetum longifolium*, are all very similar.

OPPOSITE: The flowers of *Catasetum* confused botanists completely for a long time. These plants are dioecious, meaning they have male and female flowers on different plants, or monecious, meaning they have both male and female flowers on the same plant. *Catasetum* flowers, like this *Catasetum saccatum*, are unisexual, unlike many orchids that have male and female parts in the same flower. The issue here was that the male and female flowers are so different in colour and shape that plants bearing one or the other were given different generic names. In his orchid book, Charles Darwin cited observations that hinted at the unisexuality of these plants – Richard Schomburgk observed plants that never set seed in Guyana, and a Mr Rogers of Riverhill 'informs me that he imported from Demerara a *Myanthus*, but that when it flowered a second time it was metamorphosed into a *Catasetum*'. Male flowers of *Catasetum* differ greatly between species.

CYCNOCHES EGERTONIANUM.

STANHOPEA TIGRINA.

OPPOSITE: The flowers of the bucket orchids, like this *Coryanthes speciosa*, are like the products of a fevered imagination. The bucket-like labellum gives the orchid its common name, but the strangest part is how the flower parts interact to attract euglossine bees who then pollinate the flower. The labellum is highly modified into three parts; the part closest to the pedicel (flower stalk) emits a heady fragrance, the middle part is narrowed and looks for all the world like it is the handle of the bucket-shaped lowermost part of this petal. The column, with the stigma and anthers, fits tightly along the base of this entire structure. At the base of the column are two water glands that drip fluid into the bucket-shaped part of the labellum – as male euglossines try to climb the narrowed handle to collect scent, some of them slip on its waxy surface and slide into the fluid-filled bucket. Once their wings are wet, they cannot fly out, and the walls are slippery – the only way out is through a tiny gap between the column and the bucket edge. As the bee squeezes out, he dislodges the anther cap, and the pollinaria are stuck onto his back. If the next flower he visits has already had a visitor, the stigma is exposed. So, as our bee falls into the bucket again with his cargo, he must squeeze out again, but this time the pollen masses adhere to the exposed stigma.

ABOVE LEFT: Of this plant, *Cycnoches egertonianum*, James Bateman said: 'Strange things – and no less strange than true – have already been recorded of Orchidaceous plants, but the case which is represented in the accompanying Plate casts into the shade all former frolics of this Protean tribe.' Plants gathered by Mr Skinner in Guatemala with unusual blossoms had, upon cultivation, produced just bog-standard *Cycnoches ventricosum* flowers, so a mistake was suspected. But once in cultivation the plant produced not only the *Cycnoches* flowers, but an additional inflorescence like the one Skinner had seen in the field. Bateman rejected the idea that these were male and female flowers, but he was wrong. Like most orchids with unisexual flowers, the female flowers are very similar between species, while the male flowers differ greatly.

ABOVE: James Bateman thought *Stanhopea tigrina* the most wonderful of its genus with its 'huge, fleshy lip, of so strange and fantastic a figure, that it would rather seem to have been carved of ivory, or modelled in wax, than be a bona fide production of the vegetable world.' He suggested that 'even Mrs. [Augusta] Withers's skill was taxed to the utmost to convey an adequate notion of it.' Mrs Withers was one of three female artists who drew almost all of the orchids for Bateman's book, *The Orchidaceae of Mexico and Guatemala*. She alone is responsible for 21 of the 40 plates.

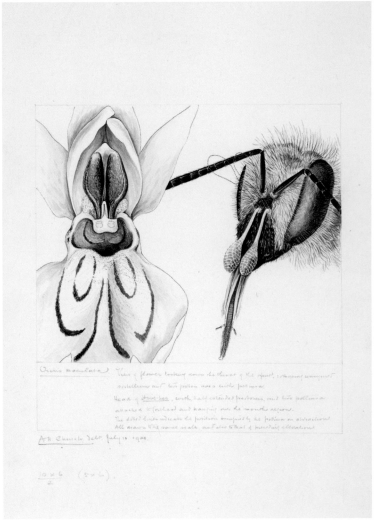

ABOVE and ABOVE RIGHT: The positioning of pollinaria on the bodies of orchid pollinators must be very precise in order for the pollen masses to hit the receptive stigma of another flower just right to achieve fertilization. Markings on the labellum, flower shape, depth of nectaries – all these contribute to lining the insect up to trigger the mechanism that releases the two pollinia with their sticky tags onto the insect body. As the sticky viscidium dries, the pollinia move into the right position to encounter the stigma in the next flower she visits. The pollinaria of the common spotted orchid, *Dactylorhiza fuchsii*, are attached to the bee's face, but other species of orchid place the pollinaria in a wide variety of other spots on insect bodies, from eyes to abdomens.

OPPOSITE: These spectacular epiphytic orchids, *Dimorphorchis lowii*, endemic to Borneo, have flowers of two colours on the same inflorescence – some with white backgrounds and red spots, the others with yellow backgrounds, each colour morph with a slightly different fragrance. Unlike those orchids with unisexual flowers though, all the flowers of *Dimorphorchis* are bisexual – they have both male and female parts. Nothing is known about the pollination of these plants – a study waiting to happen.

W. Fitch, del. et lith.

Pl. 161.

Vincent Brooks, Imp.

RIGHT: When a pollinator contacts the column upon entering the flower not much happens; it is when it leaves that the anther covering is pulled back and the pollinia, with their viscidium, attach to the insect body. In this almost abstract close-up of the column of the primrose or many-flowered dendrobium, *Dendrobium polyanthum*, the pollen masses are shown once they have detached from the column. Compare this to the colour close-up of the same orchid on page 72 to see the column before the pollinator has left.

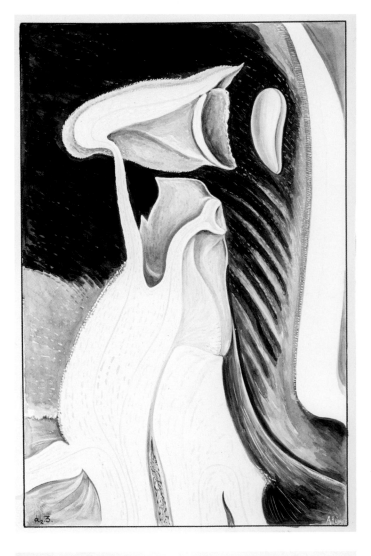

RIGHT: The German-born explorer Robert Hermann Schomburgk travelled through Guyana in the 1840s, surveying the boundaries of the then-British colony. He used a black background in his painting of this clear white orchid, *Stanhopea grandiflora*, to better show its floral structure and delicate colouring. Another of his watercolours of the same species is labelled 'River Berbice'; the Berbice is the easternmost of the main rivers flowing south to north through Guyana.

OPPOSITE: When I first saw the showy lady's slipper, *Cypripedium reginae*, near a bog in upstate New York, I was surprised to find a dead bee lodged in the flower, as if it got stuck trying to get out. Little did I know then that this was in fact what was happening. If the bee is too large to exit the space between the column and the petal, she gets stuck.

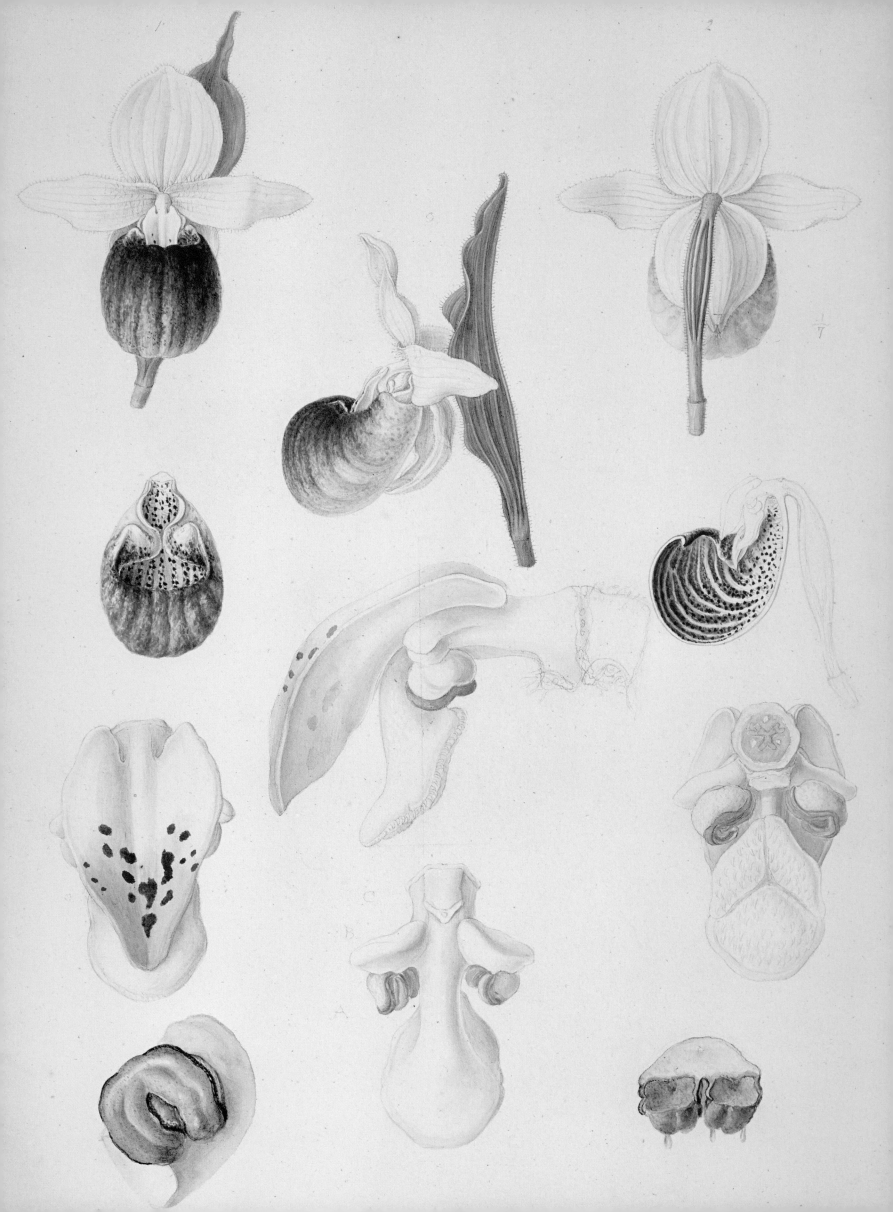

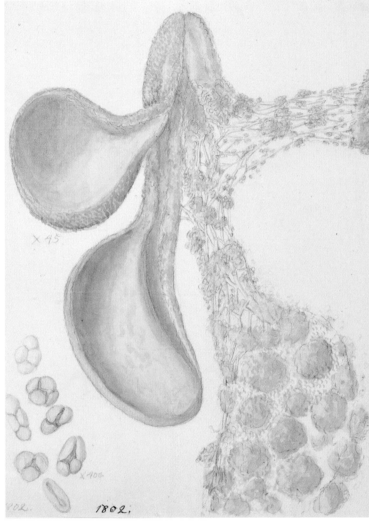

ABOVE: Although dendrobiums produce massive, sweetly scented floral displays, they offer no nectar as a reward to pollinators. The cross-section here of *Dendrobium polyanthum* shows the lack of a nectary or nectar spur, but nectar-seeking insects are guided to the centre of the flower by colour gradation that often includes a patch of high UV reflectance at the base of the bright yellow ridge along the labellum. It looks like a landing platform to a reward, but is instead a piece of false advertising.

OPPOSITE: Slipper orchids are all pollinated by deceit – they offer no reward. Pollinators of the showy lady's slipper, *Cypripedium reginae*, are medium-sized bees, who enter the flower through the hole in the labellum, but then cannot get out the same way they entered. They must exit through one of the dorsal openings – the small spaces between the column and the side petals – and in so doing they brush against the reproductive parts and either take away or deposit pollen.

ABOVE: Franz Bauer was not only a consummate artist, he was also fascinated by the mechanisms he observed though the microscope. In a paper criticizing Bauer's interpretation of the pollen structure of orchids Robert Brown, then the Keeper of Botany at the Natural History Museum in London, first described the phenomenon of what is now known as Brownian motion: '... motion of the granular fluid is seldom in one uniform circle, but frequently in several apparently independent threads or currents.' These observations of orchid pollen of *Phaius tankervilleae* also revealed an opaque area in each cell, which Brown called the areola or nucleus. Although others had observed this structure, Brown gave the nucleus its name.

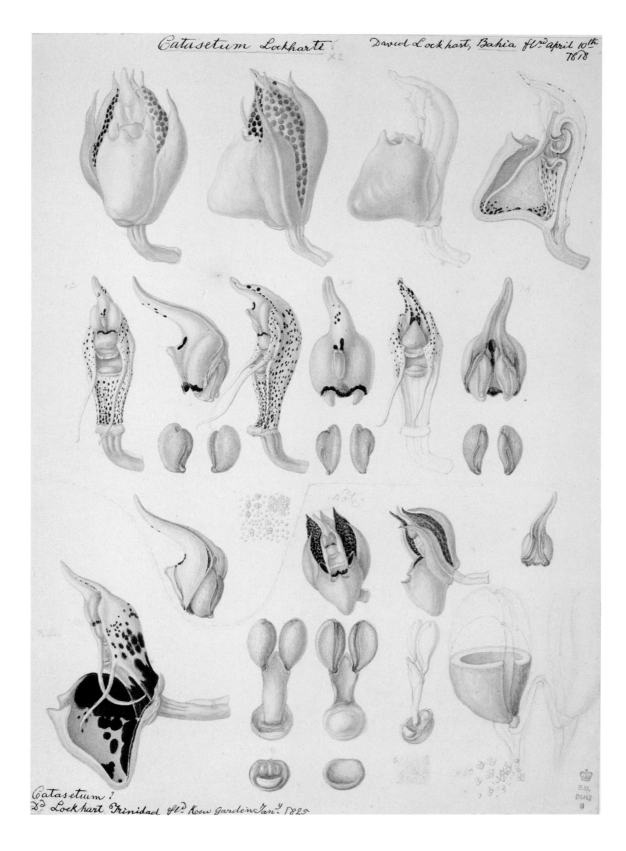

ABOVE: The male and female flowers of *Catasetum*, here *Catasetum macrocarpum*, look so different – how does pollination ever happen? Both sexes produce the same fragrance that attracts male orchid bees; the interaction of the bee with the flower though is completely different. While foraging for fragrance in a male flower the bee touches the two hair-like antennae; this triggers the movement of the column down to slam the pollinaria onto his back. This violent placement makes bees avoid male flowers subsequently, but not female flowers that emit the same scent. These flowers, however, are upside down relative to the male flowers, with the labellum uppermost, so after harvesting fragrance from female flowers the bee leaves behind the pollinia as he exits upside down.

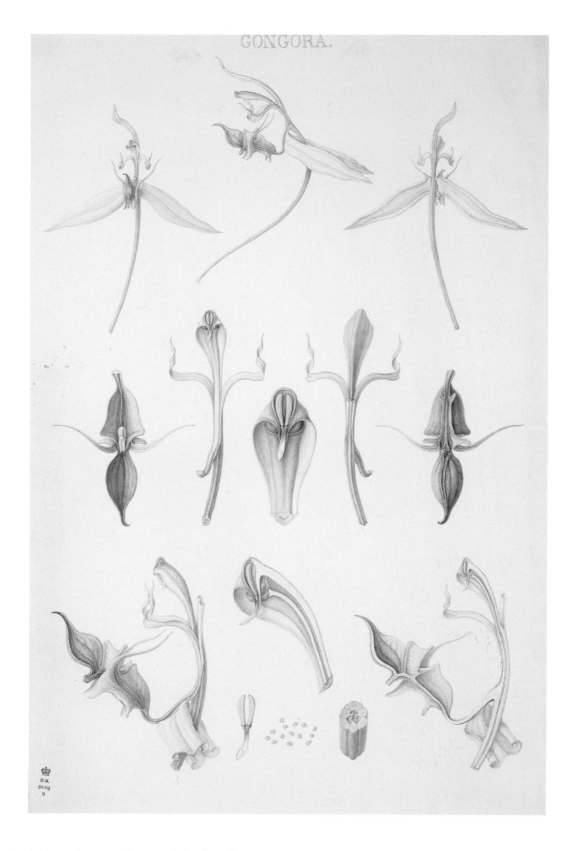

GONGORA.

ABOVE: *Gongora* flowers, like this *Gongora retrorsa*, look like something from another world. These flowers are depicted here inverted from their natural position in the wild, their colour and slightly shrivelled appearance indicate Franz Bauer painted them in monochrome wash from rehydrated dried material. The heading on the plate suggests it might have been meant for inclusion in Lindley's *Illustrations of orchidaceous plants by Francis Bauer* – only four fascicles of which were ever published due to their prohibitive cost and lack of outside sponsorship.

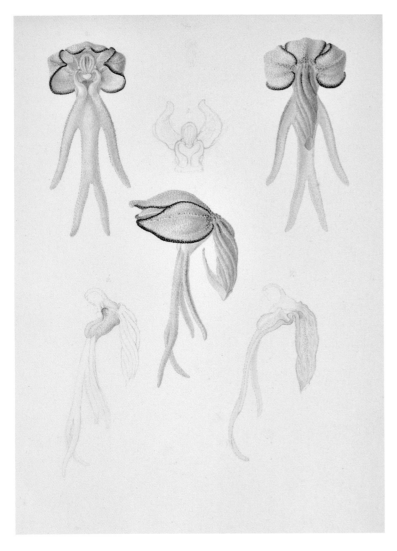

ABOVE: Orchid flowers are usually upside down, that is to say, the uppermost petal in bud, becomes lowermost once the flowers open. This petal is called the labellum, here in *Orchis anthropomorpha* shaped a bit like the outline of a person. This phenomenon is called resupination and is achieved by the twisting of the inferior ovary or pedicel of the flower through 180°. Some flowers though are right side up, with the uppermost petal in bud still uppermost in flower. But rather than just not twisting at all, these flowers usually twist through 360°, something Darwin cited as compelling evidence against special design.

ABOVE: Many species of sun orchids of Australia and New Zealand, like this *Theymitra venosa*, are an unusual colour for an orchid – bright clear blue. They have very regular flowers, compared to most orchids, but the column is often brightly contrasting and yellow. It is thought that the sun orchid flowers are mimicking buzz-pollinated flowers of other plants in the same habitat that offer pollen, not nectar, as a reward. The models for these mimics have poricidal anthers that a bee must vibrate with indirect flight muscles to release pollen (buzz pollination). The same behaviour on the orchid flower attaches the pollinaria to bees visiting in the hopes of a pollen reward.

OPPOSITE: *Anguloa* species like this *Anguloa × ruckeri* are known as tulip orchids, for their enlarged cupping sepals. Like other Neotropical orchids pollinated by orchid bees, they produce a cocktail of fragrance compounds. The main ones produced by some tulip orchids are alpha-pinene and cineole – these terpenes are also parts of pine, rosemary and eucalyptus odours.

Nugent Fitch lith.

ANGULOA RUCKERII SANGUINEA.

J. Nugent Fitch imp.

no. 13
perfumed, a little like lily of the valley

very common on rocks all through the district

6

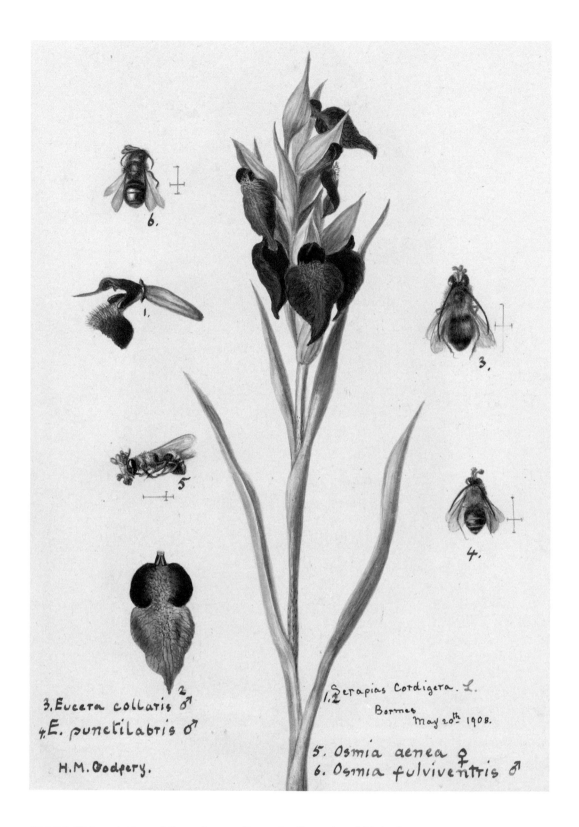

ABOVE: This painting of *Serapias cordigera* by the natural historian Hilda Godfery shows the positions of pollinaria on the bees she found visiting the flowers of the heart-shaped serapias in Bormes-les-Mimosas in the Côte d'Azur in May of 1908. Both male and female bees visit these flowers and shelter in the cavity provided by the arched sepals – no other reward is offered.

OPPOSITE: The perfume of *Angraecum sesquipedale* flowers is produced in the evening and is variously described as 'very spicy, masculine and sometimes overpowering' and 'like that of a garden lily, *Lilium candidum*', jasmine-like and pleasant, but strong enough to fill a room with scent. The Reverend William Deans Cowan likened it to 'lily of the valley' and recorded this plant as very common – what the forest must have smelled like! Charles Darwin suggested that the flowers must be pollinated by a 'gigantic moth'; in fact, it is the proboscis of the moth *Xanthopan morganii praedicta* that is gigantic, the moth's body is of a normal size.

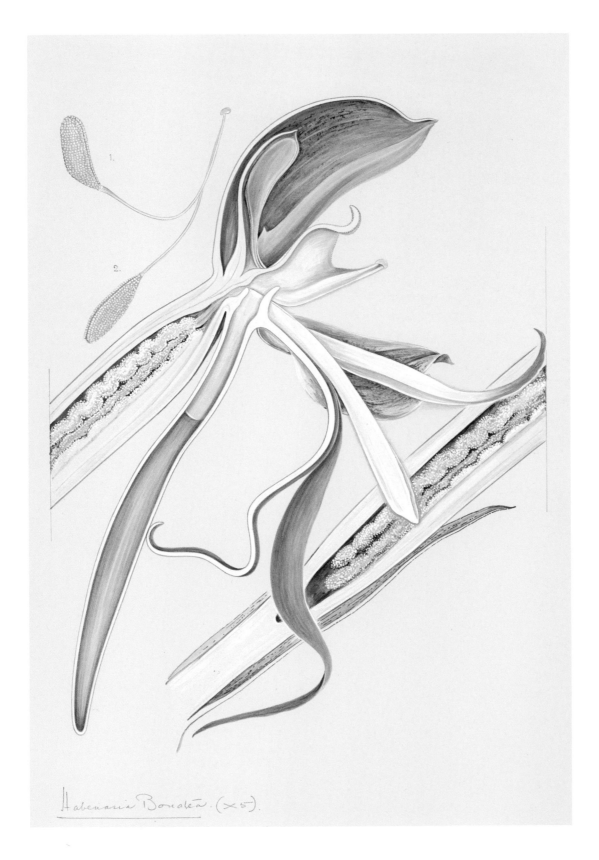

Habenaria Bonatea. (×5).

ABOVE AND OPPOSITE: This South African orchid, *Bonatea speciosa*, is called moederkappie in Afrikaans, in reference to its green and white, bonnet-shaped flowers. The flowers begin to produce scent and are visited by hawkmoths for just a short window right at dusk. So far, so simple. But the pollination of this orchid is extraordinary, even in this most extraordinary of families. As the moths probe for sugar-rich nectar in the spur they must go to one or the other side of the little tooth-like projection at the entrance, so their heads turn ever so slightly and only one of the pollinaria attaches to the lower part of the moth eye. The pollinaria are elongated, so once attached they dangle from the moth's eye. When the next flower is visited the pollinaria are channelled by the curved elongate green petal lobes and brush against the stigma – the long white horns at the centre of the flower. The removal of one pollinarium at a time by the moths is an interesting reversal from the all-at-once pollen removal of many other orchids.

Interdependence

'The outstanding characteristic of the Orchids that is regarded as a sign of decadence is lavish yield of seed coupled with sparse distribution.'

Oakes Ames, 1922

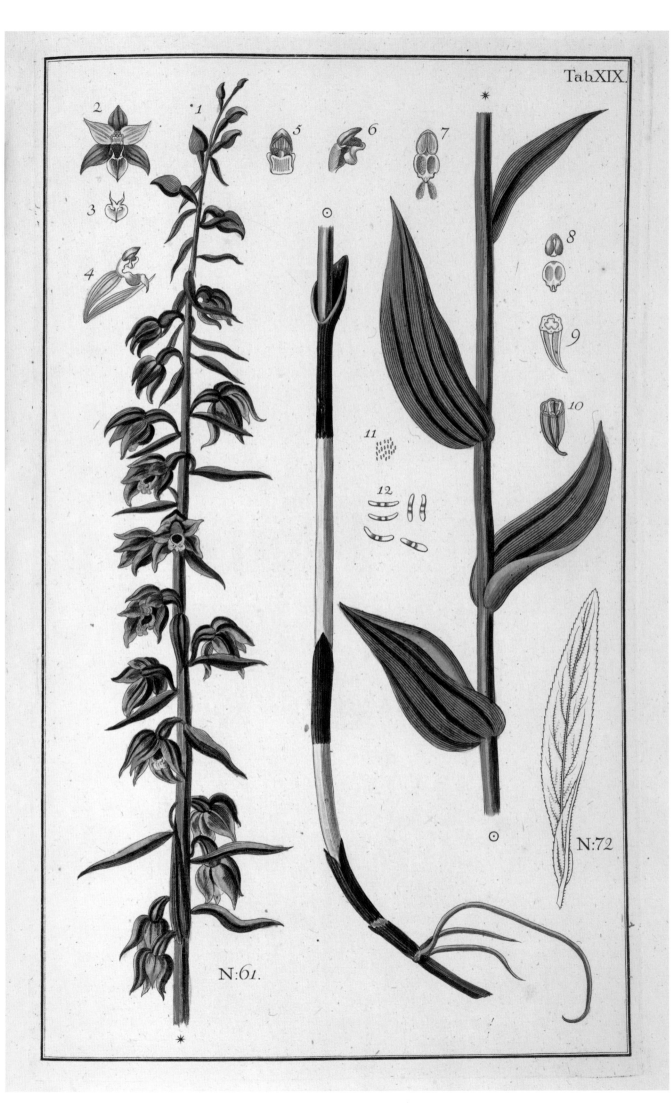

Tab.XIX.

N:61.

N:72

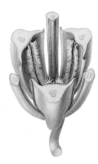

O RCHID REPRODUCTION WAS mysterious to early botanists. Where were the seeds from which the next generation of plants grew? Hieronymus Bock, whose *Kreütterbuch* of 1539 was written in his vernacular German rather than Latin, noticed that a dust was produced from where the flowers had been, but he did not connect this with the generation of new orchids. About a hundred years later, Georg Everhard Rumpf (or Rumphius), a Dutch merchant in Ambon, one of the Indonesian Maluku Islands, described orchid seeds as 'yellow meal' or 'sandy lint' and suggested they might be dispersed by the wind – 'it is likely, that they are either seeded from their own seed by the wind, or by birds'. A crude drawing of helleborine orchid seed appeared in Konrad Gessner's posthumously published *Opera Botanica* of 1771, looking like sausages with two tiny dots inside. Orchid seeds are indeed tiny; they are the smallest seeds in the flowering plants and appear dust-like to the naked eye. No wonder medieval herbalists thought orchids reproduced by spontaneous generation.

As with all flowering plants, once the flower is pollinated, the ovary swells to become a fruit that contains the seeds. Angiosperms – or flowering plants – literally means 'hidden seeds', so perhaps we should call them fruiting plants and not flowering plants? Orchid fruits are capsules made up from the three carpels, or ovule-holding structures, in the centre of the flower. Capsules are dry fruits that split at maturity to release the seeds; capsular fruits are usually brown, dry and nondescript, not the sort of things one might want to eat. One exception that we have all eaten, but probably didn't realise it, is vanilla.

Vanilla beans are not beans at all – they are orchid capsules. The name vanilla comes from the Spanish common name – *vainilla* – meaning 'little sheath' or 'little pod'. The Aztec name for this plant was *tlilxóchitl*, or 'black flower' referring, not to the flowers at all, which are greenish white, but to the capsules that turn black once taken from the plant and dried. The best vanilla comes from the Mexican species *Vanilla planifolia* – the plant the Aztec peoples used, now also grown commercially in Madagascar and known as Bourbon vanilla. The vanilla production process is incredibly labour-intensive. The flowers are pollinated by hand, capsules are harvested by hand, and preparation of the pods involves a prolonged process of sweating (fermentation), drying and ageing to maximize the aromatic vanillin content in the resulting product.

OPPOSITE: *Epipactis palustris*. Published in the sixteenth century, long after the author's death in 1565 from plague, Konrad Gessner's botanical works were careful, meticulous observations of nature. His illustrations to 'help the reader come more easily to the knowledge of unfamiliar things' were based on his own drawings and were among the first to consistently use details of plant parts.

The pulpy 'caviar' in the centre of the dried black pods, composed of pulp and tiny seeds, is then used to make vanilla extract or directly used in ice cream and other manufactured products. The flecks of brownish black you see in expensive vanilla ice cream could be tiny orchid seeds but is more likely to be fragments of pod – all part of the orchid capsule and its thousands of tiny seeds.

In the late 1960s, ecologists Robert H. MacArthur and Edward O. Wilson characterized organisms along a spectrum depending upon the quantity and quality of offspring they produced. 'k-selected' species produced few, expensive offspring,

whereas 'r-selected' species produced many, cheap offspring; k-selected organisms were thought to live in stable environments, care for their young and be long-lived, whereas r-strategists occupied unstable environments, left their young to fend for themselves and were short-lived. This paradigm has largely been replaced with models more supported by real-world data, but orchids, with their huge numbers of tiny seeds, certainly produce young in quantity. Seeds of most plants contain within them tissue that serves to nourish the growing plantlet as it gains size and ability to photosynthesize for itself; familiar examples of this are the cotyledons of beans or the endosperm of maize – both of which also serve humans as food. Orchid seeds are different. They consist of merely a tiny embryo surrounded by a flimsy, almost transparent layer of cells – no nutritive tissue and no protection at all.

Although orchid seeds had been known for more than a century, nobody had observed

ABOVE: *Vanilla planifolia*. Writing of tlilxóchitl, or the black flower of the Aztecs, the sixteenth century Spanish naturalist and recorder of Aztec plant lore Francisco Hernández said: 'This plant is voluble, with leaves like the plantain but fleshier and longer. They are dark green, sprouting at intervals along the stem. Long, narrow sheaths that are almost round grow from the stem. These sheaths, the vanilla beans, smell like musk or balsam of the Indies, and they are black, hence the name. It grows in hot moist places. ..... also said to be one of the most aromatic plants of the region.'

them germinating until the early nineteenth century, and then by accident. Seedlings were seen in the field in Britain in 1802, but the rather unprepossessing and tiny-flowered Brazilian species *Prescottia plantaginea* germinated freely in greenhouses in London around the middle of the nineteenth century. It was not a big hit, though, not being one of the showier orchids around. But because orchid seeds were so difficult to germinate, any seedlings seen made the news, and growers tried all sorts of methods to grow orchids from seed. They tried using fine soil covered with moss, or blocks of wood covered with moss – nothing worked consistently. In 1849 David Moore, director of the Glasnevin Botanic Garden in Dublin, reported the germination of several species of epiphytic orchids, some of which grew on to flower; this was the first reliable report of orchid seed germination that was not by chance.

It was not until the turn of the nineteenth century that the key to unlocking orchid seed germination was found – French botanist Noel Bernard discovered that fungi known as mycorrhizal fungi were necessary for the successful germination of orchid seeds. Mycorrhizal fungi had been known since the middle part of the nineteenth century, and their relationship with plants was beginning to be understood. These fungi colonize the root system of a host plant and provide water and nutrient absorption capacity. The plant, in turn, provides the fungus with the carbohydrates produced by photosynthesis. This mutually beneficial relationship is a

ABOVE: The seeds of this epiphytic Neotropical epidendrum, *Epidendrum secundum*, clearly show the elongate tails that help lodge the embryo in moss on branches.

symbiosis, or mutualism. Mycorrhizae (the symbiotic association of the plant and fungus) are found in an estimated 85 per cent of all land plant species, from moss relatives to giant forest trees. In fact, forests are often connected by complex nets of mycorrhizae that have been termed the 'wood-wide web'; mycorrhizae have even been implicated in plant-to-plant communication and stress responses. This is an ancient association; mycorrhizae have been found in the fossils of the Rhynie Chert from the Devonian period (400-412 million years ago) and are thought to have been crucial to the transition of plants to land. Even tiny plants such as bryophytes have mycorrhizae – both liverworts and hornworts have them, but mosses do not; it is not known whether this is a loss from a mycorrhizal state or whether mosses never developed these partnerships. One could say mycorrhizae are almost ubiquitous.

Historically, mycorrhizae were divided into two basic sorts – endomycorrhizae, in which the fungal hyphae (filamentous structures that are the asexual bodies of most fungi) are found inside plant tissues, and ectomycorrhizae, in which the hyphae form a sheath around the host root but do not invade the cells themselves. Since then, this simple distinction has been extended, and the endomycorrhizae have been subdivided into several groups, one of which is found only in the orchids. Orchid symbioses involve at least two major orders of fungi. The fungal hyphae form coiled, ball-like structures known as peletons, unique to orchid mycorrhizae. Because the fungal partners in orchid mycorrhizae are in the asexual state, identification can be very difficult, and many of these fungi are lumped into a form-taxon called *Rhizoctonia*. The ability to sequence DNA has enabled better understanding of the diversity of fungi involved with orchids, but it can be difficult to know which fungi are obligate associates and which are just free-living fungi from the surrounding environment.

When seen close up, orchid seeds are peculiar. The tiny embryo is enclosed in a sheath-like structure that is a single cell layer thick and usually open at one end. Some even have coiled thread-like cells at the tips that unravel when wet and fix the seed to moist bark or moss. But without the nutrition produced by a fungal partner, the seed is destined for death – for germination and ultimate survival it must be infected by the right sort of fungus. The mycorrhizal fungus provides food – the very thing the flimsy orchid embryo lacks. Once an orchid seed lands in a suitable place where it can take up water, it begins to germinate by producing a few root hairs. If the right fungus is present, its hyphae quickly colonize the root hairs; this stage is known as the protocorm. Eventually the orchid produces leaves that can carry out photosynthesis and provide the food needed by the young orchid. Many seeds must fail to be infected with an appropriate fungus — a good reason to have many small seeds with exceptional dispersal ability. That way, at least some will make it to adulthood.

Many small seeds, most of which perish, is a fine strategy for a plant in the wild, but orchid fanciers wanted to germinate the results of cross-pollinations and grow rare and wonderful orchids from seed. The first artificial hybrid orchids had been made in the late-nineteenth century by nurseryman H. J. Veitch and his orchid grower John Dominy, via hand pollination. When they showed their 'Orchidaceous mule' to the orchid specialist John Lindley he exclaimed 'Why, you will drive the botanists mad!' But seeds of these specialities and of many of the prized tropical orchids were difficult to germinate and grow; it took the skill of specialist growers like Veitch and Dominy to make it happen. Orchid hybridization and cultivation were not for everyone.

Things were about to change, however. In the early twentieth century American botanist Lewis Knudson realized that the orchid–fungus relationship at the seedling stage was all about food – his interests were not so much in orchids, but in the enzymes produced by fungi and the use of carbohydrates and amino acids by plants. He developed artificial media with added sucrose and other compounds – and the seeds grew in the absence of any fungal association! Knudson thought the fungi were not necessary for

ABOVE: The capsules of most terrestrial orchids, like this *Cypripedium acaule*, are held upright, even if the flowers are not. This allows space for them to shake and blow in the slightest breeze to release the tiny seeds.

J.C.Davy, pinxit

MDCCCXII.

orchid germination; arguments raged between him and a series of orchid biologists, one of whom famously said '.... An orchid seedling without its fungus is like Hamlet without the Prince of Denmark.' These detractors all maintained that the fungus–orchid relationship was a specific and important one while Knudson pooh-poohed the idea. It turns out that both sides were probably right. You can germinate and grow orchids on media without fungal symbionts, but many of them do better with a fungal partner.

Some orchids never photosynthesize; they retain their fungal symbionts their entire lives, relying on the fungus for carbon, amino acids and minerals. These sorts of plants used to be described as saprophytes but are now more correctly called obligate mycoheterotrophs, reflecting the fungal relationship involved and their lack of the chlorophyll molecules needed for photosynthesis. This is a parasitic, rather than a mutualistic, relationship – the orchids unilaterally depend upon the mycorrhizal fungi for their nutritional needs.

These pale ghostly plants are often found on the forest floor among the leaf litter and generally pass unnoticed to those not seeking them. The British ghost orchid, *Epipogium aphyllum*, is keenly sought by plant twitchers, but it eludes them. In 2018 two enthusiasts who independently travelled the country to see all British orchid species both found them all – except the ghost. The last confirmed sighting is from 2009 and the one before that was in 1971, but people keep on looking; there is even a Twitter account devoted to the search for this little elusive plant. It seems to pop up here and there, and may be more common underground than we think, only emerging to flower every now and then.

ABOVE: The tiny delicate ghost orchid, *Epipogium aphyllum*, probably spends most of its life underground, so the more orchid hunters trample and compact the soil in forests looking for it, the less likely it might be to come up and flower!

OPPOSITE: Coralroot orchids, like this *Corallorhiza maculata*, take their name from the intricately branched underground stem system with a huge area for plant-fungus interaction and infection, necessary for these mycoheterotrophs.

Corallorrhiza maculata Raf.
Large Coral-root

M.E.Eaton

Large Coral-root

Not all mycoheterotrophs are tiny and elusive, however. Some tropical climbing mycoheterotrophic orchids can be tens of metres long and climb high into the canopy; these must be the largest mycoheterotrophs in the world. Roots of obligate mycoheterotrophs are peculiar – they are short and stubby, maximizing the area for fungal hyphae penetration.

Initial studies of achlorophyllous (lacking chlorophyll) mycoheterotrophs suggested that they depend largely on a single mycorrhizal fungal species, but subsequent work has shown that the variation in fungi is as great as that of the orchids themselves. For some temperate-forest orchids such as *Corallorhiza*, the distribution of the orchid is highly dependent on the abundance of the fungal partner, indicating high specificity, whereas others, such as some species of *Cephalanthera*, accept a wide variety of fungal partners independent of their abundance in the environment. In a seemingly peculiar twist, the carbon passed from the fungus to the orchid is derived from nearby photosynthetic plants, making many of these orchids what are known as epiparasites – parasites of photosynthetic plants by their fungal proxy. The mycorrhizal fungi are in a mutualistic relationship with the photosynthetic partner, providing soil minerals in exchange for carbon, and the orchid takes carbon, amino acids and minerals from the fungus.

In other orchids, especially in the tropics, specificity can be different. Many orchids that have been tested have exhibited larger communities of fungal partners; they are apparently less fussy about the fungus with which they associate. Association with wood-rotting fungi is a general pattern in all orchids, especially in the wet tropics. The

ABOVE: In New Zealand this species, *Gastrodia sesamoides*, is known as the pot-bellied orchid, but in Australia it is called cinnamon orchid, in reference to the strong cinnamon scent of the flowers. It was described by the botanist Robert Brown from collections made at the turn of the nineteenth century voyage of HMS *Investigator*, captained by Matthew Flinders. Ferdinand Bauer was the artist on the voyage, painting the extraordinary plants seen in the southern hemisphere.

fungi obtain their nutrition from dead and decaying organic matter, such as wood, and not from living photosynthetic plants; these fungi are known as saprophytes. Their hyphae forcibly penetrate most solid materials and then they enzymatically digest them; orchids then benefit from the digestive power of these fungi, from which they receive all the nutrition they need. Botanist William Stearn characterized the orchid–fungus relationship in colourful terms: 'From Karl Marx's standpoint, wealthy Victorian orchid-growers enjoyed their orchids as a consequence of the sweated labour of underpaid miners that they never saw. Research on mycorrhiza suggests that orchids can be regarded in much the same way – as ostentatious floral capitalists dependent upon the obscure activities of fungi.'

The complexity and diversity of orchid–fungus relationships in both green and achlorophyllous orchids mean that the tiny seed of an orchid, once released from the capsule, has a lot to contend with before it survives to grow and reproduce. Not only must the right fungal partner be present in the environment it lands in, but the habitat itself must be suitable as well: light, water, other plants – all these make a difference. The ability to germinate orchid seeds *in vitro,* on media containing the right fungus to provide food, is a step to growing orchids commercially for sale by the thousands in garden centres and supermarkets. But leveraging the knowledge of orchid germination and fungal relationships is even more important for orchid conservation. Scientists looking to establish or re-establish populations of native orchids have learned to think like orchids themselves, taking into account all the complex interactions and interrelationships that are involved in orchid biology. Each orchid conservation case is different and requires exploration of the germination story each time – this may seem frustrating, that there is no simple method, but nature is full of complexities that we still do not fully understand.

Fitch del et lith.

Cattleya Mofsiæ.

Reeve imp.

Plate 29.

ABOVE: The orchids that young people of my generation gave to each other as decorations for the Senior Prom were usually cattleyas, like this *Cattleya mossiae*. Big, showy and beautiful – they are the orchids of romance. But *Cattleya* was found by accident amongst the moss and lichen packing another shipment and grown on by the nurseryman William Cattley. The genus was named in his honour by the orchid specialist John Lindley.

OPPOSITE: *Phalaenopsis*, or moth orchids, are among the most commonly sold supermarket or grocery store orchids. Most plants we see for sale are unnamed hybrids, rather than the species, *Phalaenopsis lowii*, depicted here. Epiphytic in their natural habitat in Southeast Asia, this orchid is monopodial, with single, rather than multiple, branches of blossoms per stem. They flower at any time of the year – just wait and yours will bloom again!

Pl. 168.

W. Fitch, del. et lith.

Vincent Brooks, Imp.

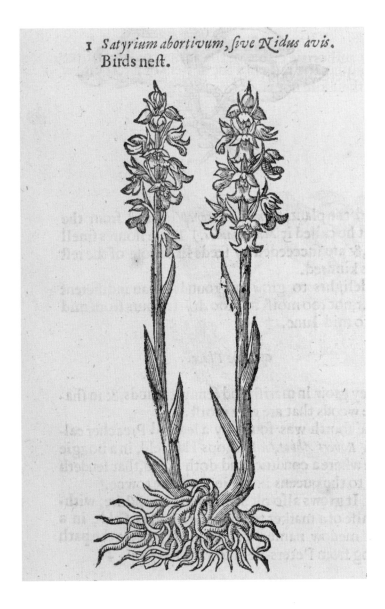

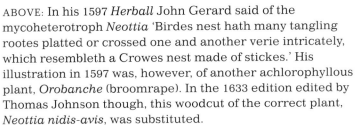

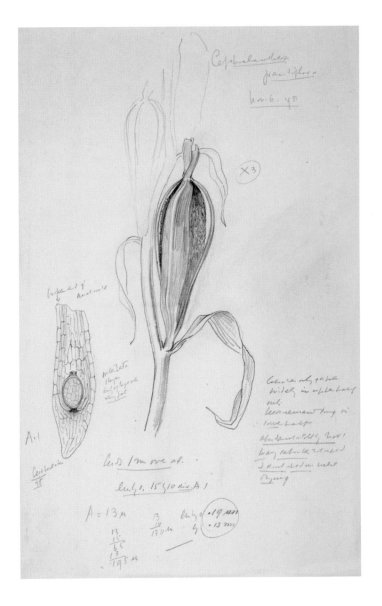

ABOVE: In his 1597 *Herball* John Gerard said of the mycoheterotroph *Neottia* 'Birdes nest hath many tangling rootes platted or crossed one and another verie intricately, which resembleth a Crowes nest made of stickes.' His illustration in 1597 was, however, of another achlorophyllous plant, *Orobanche* (broomrape). In the 1633 edition edited by Thomas Johnson though, this woodcut of the correct plant, *Neottia nidis-avis*, was substituted.

ABOVE: The exquisitely detailed flower paintings by Arthur Harry Church are considered some of the finest botanical illustrations of the twentieth century. His attention to detail and technique of painting on Bristol board make the lines and structures of each subject crisp and clear. This sketch of a sword-leaved helleborine, *Cephalanthera longifolia*, shows his attention to detail went beyond flowers to all plant structures – his notes on this image include measurements of seeds and embryo, as well as comments on the opening of the orchid capsule.

OPPOSITE: The seeds of the violet limodore or violet bird's nest orchid, *Limodorum abortivum*, are among the largest in the orchid family and the plants are very slow growing. Seedlings can remain completely underground for eight to ten years, nourished solely by their associated mycorrhizal fungi. Plants only emerge above ground to flower; their purple flowers are more striking that those of many other European mycoheterotroph orchids.

T. 193.

Orchis abortiva

Pl. 13.

Mrs. Withers, del.   W. Gauci, lith.

CATTLEYA SKINNERI.

Pub.d by J. Ridgway & Sons, 152, Piccadilly, Sept.r 1.st 1838.

Printed by P. Gauci, 9, North Crescent Bedford Sq.re

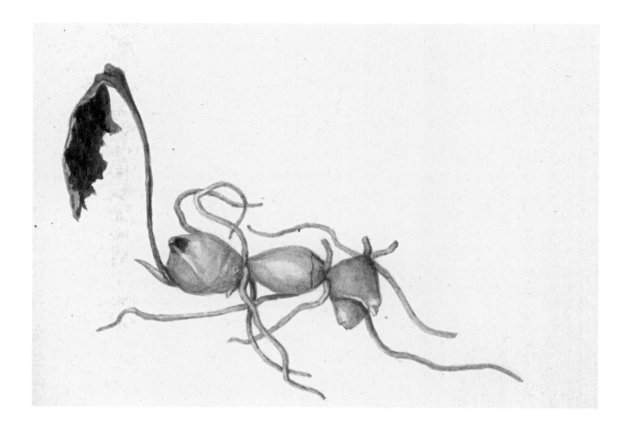

ABOVE: The crane-fly orchid, *Tipularia discolor*, is an extremely peculiar plant. Its underground stems are starchy, edible and potato-like. The plants bear a single leaf each autumn, which persists over the winter, and dies back in the spring, when the plant flowers. The flowers are twisted sideways and oriented so the pollinaria are deposited on the eyes of visiting moths, the pollinators of this species.

OPPOSITE: *Cattleya skinneri*. Once Lewis Knudson figured out how to grow orchids en masse on artificial media, the orchid industry took off. No longer the preserve of rich collectors, orchid flowers were for everyone. By the 1930s the American orchid grower Thomas Young had amassed a business worth millions of dollars; ironically at the height of the Great Depression, the sale of flowers boomed for those who could afford them. Young's nurseries were characterized as the 'General Motors of the orchid business'.

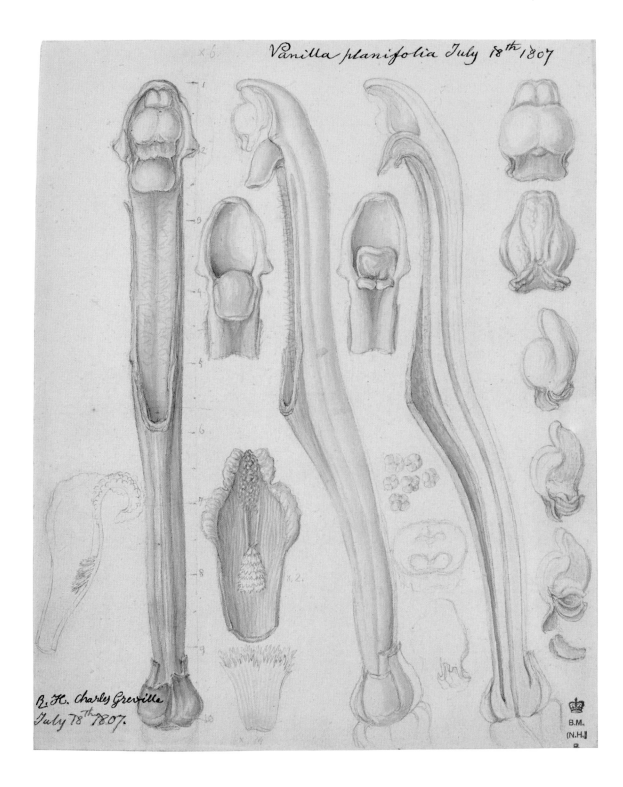

ABOVE AND OPPOSITE: *Vanilla planifolia*. The seeds of vanilla are unusual in the
orchid family. They are still tiny, but rather than being encased in a loose sheath, are
dark coloured, hard and round. Vanilla seeds are probably dispersed by animals
eating the relatively fleshy fruit; to make commercial vanilla the fruits are picked
before maturity. When vanilla was introduced to the island of Réunion (Île Bourbon)
it never set fruit, until an enslaved boy named Edmond Albius found that fruits resulted
when male and female parts of the flower were pressed together. The orchid bees
that visit vanilla flowers in their native habitat do not occur on Réunion, or anywhere
outside the American tropics. Once outside their native forests, vanilla plants did
not set fruit reliably, and so were not a particularly good prospect commercially.
With Edmond's discovery the Bourbon vanilla industry was born, and the French
broke the Spanish monopoly on the sale of that most valuable spice, vanilla.

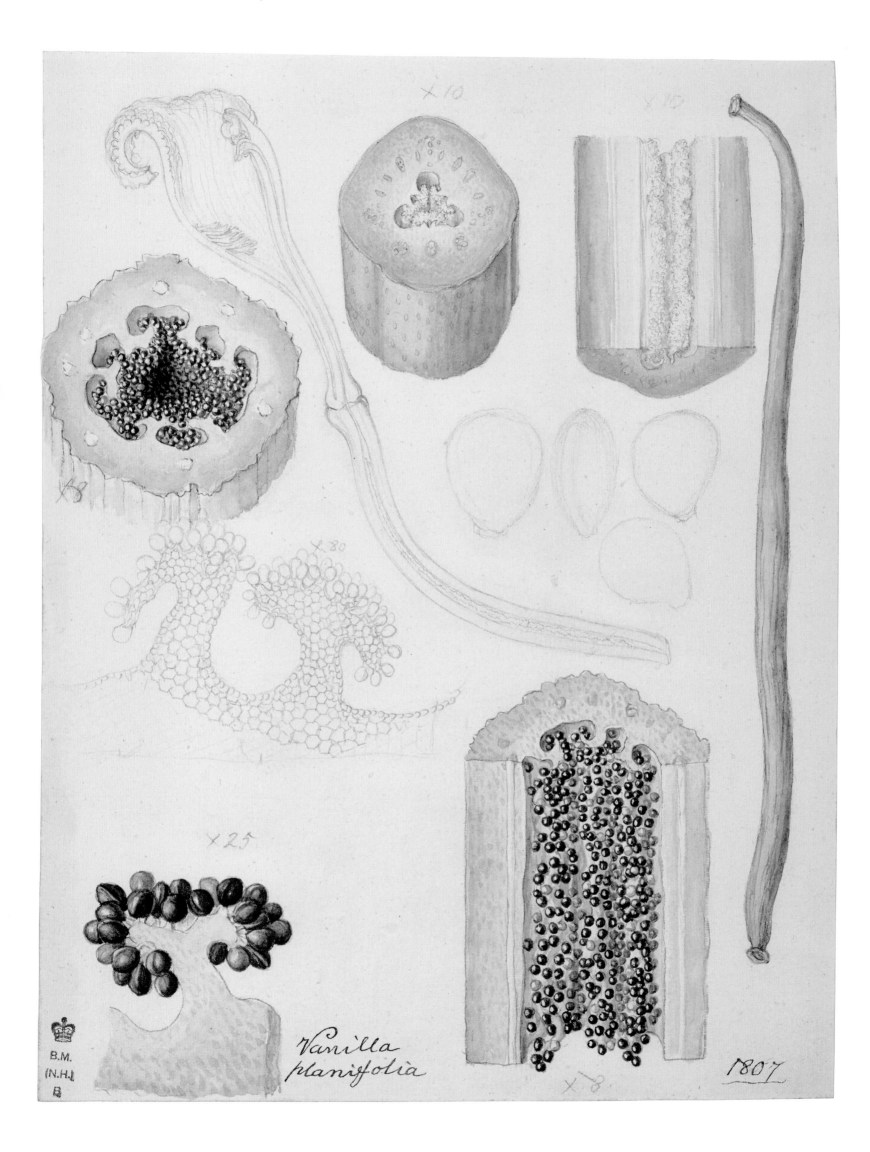

×10

×10

×80

×25

B.M.
(N.H.)

*Vanilla
planifolia*

×8

1807

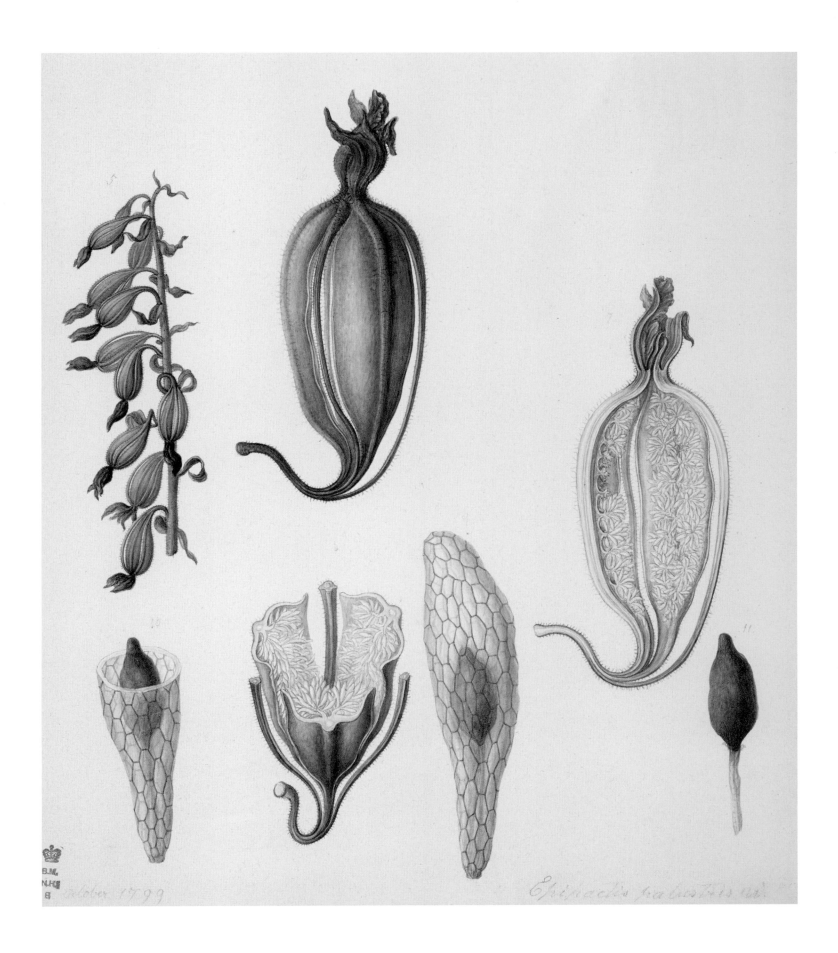

Epipactis palustris

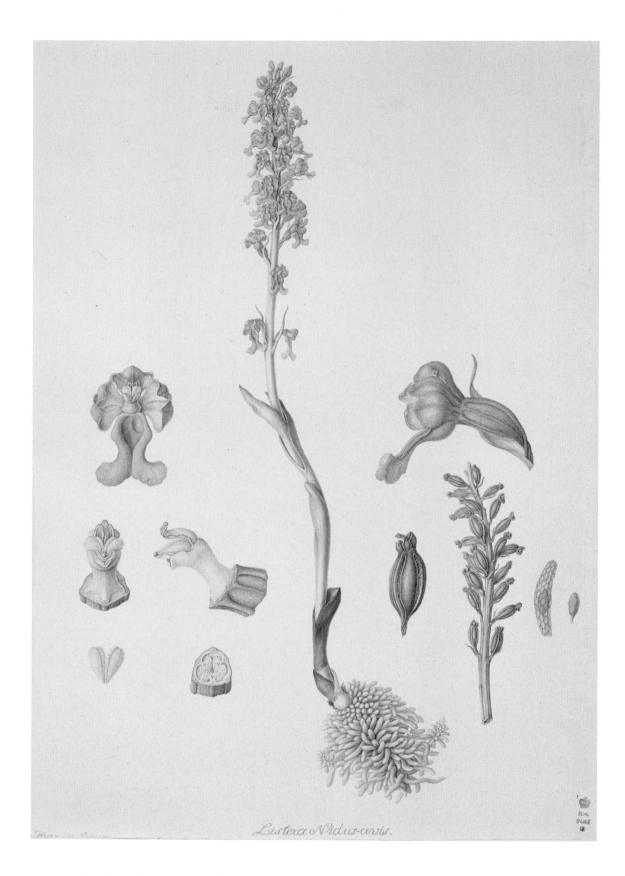

Listera Nidus-avis.

ABOVE: *Neottia nidis-avis*. Orchids without chlorophyll were often assigned to their own genus because they looked so different to plants with leaves and chlorophyll. The genus *Neottia* used to be reserved for mycoheterotrophic plants, but studies using DNA sequences showed that these achlorophyllous plants were part of a larger group of photosynthetic orchids, the twayblades.

OPPOSITE: The marsh helleborine, *Epipactis palustris*, has green leaves, thus obtaining some of its nutrition from photosynthesis. But these plants show that the dividing line between autotrophs and mycoheterotrophs in the orchid world is anything but clear. Helleborines also have dependent relationships with fungi from which they receive nutrients. The marsh helleborine is specialized compared to other helleborines and relies particularly on a single type of ascomycete fungus for this relationship.

ABOVE: The flowers of the European early coralroot orchid have the typical orchid morphology. The details in this painting of *Corallorhiza trifida* by Franz Bauer clearly show the resupinate nature of the flowers – the ovary is twisted so the petal that is uppermost in bud becomes lowermost in flower. This petal is called the lip or labellum and is often patterned or spotted.

OPPOSITE: This European early coralroot orchid, *Corallorhiza trifida*, is the only species in the genus that does any of its own photosynthesis – it doesn't have any leaves to speak of, but the pale green flowers have chlorophyll and can fix carbon dioxide. The plant depends on its associated mycorrhizal fungus for most of its nutrition.

RIGHT: *Phaius tankervillae.* Most orchid capsules do not open up completely; instead the valves remain attached at the top, making a structure that functions like a pepper shaker. Seeds must sift out through the openings which themselves are sometimes attached edge to edge by fibres, creating a sieve through which seeds are released to be blown to their destination.

OPPOSITE: Autumn lady's-tresses, like this *Spiranthes spiralis,* are easily recognized by the spiral arrangement of their flowers, flowering in early autumn after the leaves die back. The tiny seed does not travel far, so large populations can establish in the right circumstances. Uncut lawns are one such place – I was surprised to see hundreds of autumn lady's-tresses in a lawn in Kent, UK, a veritable carpet of orchids.

RIGHT: In 1842, the great Victorian orchidologist John Lindley named this orchid *Coelia baueriana* in honour of his close friend and fellow orchid-lover Franz Bauer, whose detailed paintings brought the world of orchid structure to life. Sadly, James Edward Smith had previously described the same plant as *Epidendrum tripterum* in 1793. Under the rules of botanical naming, the species name *triptera* has priority over that honouring Bauer, so we must use the name *Coelia triptera* instead.

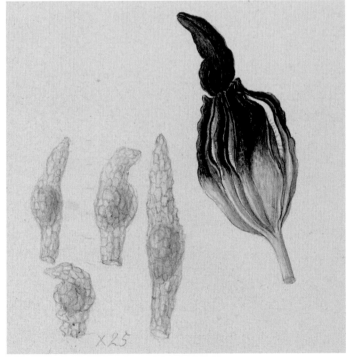

J. C. Darby, pinxit
MDCCCCVI.

ABOVE: The tiny delicate plants of the lesser or heart-leaved twayblade, *Neottia cordata*, are often found growing in thick cushions of moss in acidic habitats, they are easy to miss in dense cushions of sphagnum. Despite its apparent rarity, this orchid is distributed all around the northern part of the northern hemisphere, and even occurs on southern Greenland. Like other species of *Neottia*, it is at least partially mycoheterotrophic as a mature plant, as well as when a seedling.

OPPOSITE: *Neottia ovata*. The similarities in flower form and detailed structure led some early botanists to combine the species of the genus *Listera*, in which this twayblade orchid was long placed, with the mycoheterotrophs of *Neottia*. This was largely ignored until DNA sequence data strongly supported this decision. The many differences in vegetative form can be attributed to the acquisition of the mycoheterotrophic lifestyle; as one author put it 'acquisition of mycoheterotrophy is not in itself sufficient justification for recognition as a genus.'

Cunning deceivers

'[*Ophrys insectifera*] flowers bear such a resemblance
to flies that an uneducated person who sees them
might well believe that two or three flies were
sitting on the stalk. Nature has made
a better imitation than any art
could ever perform.'

Carl Linnaeus, 1745, *Öländska och Gothländska Resa*
(Öland and Gotland Journey),
translated by Åsberg & Stearn, 1973

PLANTS LURE THEIR POLLINATORS by the offer of rewards – often nectar or pollen, but sometimes oil or, in the case of many orchids, volatile chemical compounds used in attracting mates. In all these interactions, the insect receives some benefit from the visit, and they do not go away empty-handed. However, many plants trick insects to visit the flower with the promise of a reward but no follow-through. The uncanny resemblance of some orchid flowers to insects or mushrooms is not just a quirk of development or a case of mistaken identity due to an overactive human imagination – this resemblance is at the heart of a deceitful, almost certainly manipulative, pollination system.

The dark foreboding flowers and strange filamentous sepals of many *Dracula* flowers are not the only peculiar thing about this orchid genus, which was named in the late 1970s by orchidologist Carlyle Luer from the Latin *dracula*, 'a little dragon' (*draco* meaning 'dragon'). This is an allusion to the fancied appearance of many of the species. The reddish-brown flowers might also be suggestive of Count Dracula of vampire novel fame. These orchids have three large fused sepals with long, filamentous tips, and the labellum (lower petal or lip) is often cup-shaped with raised ridges inside. If you can get beyond thinking the flowers look like impish faces, you can see the resemblance to the underside of a gilled mushroom in the lip. People had speculated that the flowers attracted fungus flies, which normally lay eggs in mushrooms, but instead, after an attempt to lay eggs in the flowers, these deceived flies then carry away pollinia. It turns out this is largely true, but even more complicated than was thought. An Ecuadorian species, *Dracula lefleurii*, attracts drosophilid flies (fruit flies allied to those commonly used in lab experiments) by both smell and sight – the spotted sepals fool the flies into thinking there are other flies there. As a fly moves about laying eggs, she makes contact with the pollinia, which then stick to her body. Then when the fly visits another flower, the pollinia are deposited on the stigma. Sadly, the eggs never hatch – a *Dracula* flower is no good for fly larvae A relationship that is great for the orchid, is not so good for the flies. But it is not all bad; in *Dracula* flowers these flies find others and mate, and it is thought that yeast growing on the flower surface may serve as a food reward to the visiting flies.

Charles Darwin would have loved *Dracula* – but he didn't describe these flowers in his 1862 orchid book. Even in their previous classification as members of the genus

OPPOSITE: *Dracula chimaera*. It is easy to see why these orchids were named *Dracula*, the flowers almost leer at the onlooker.

Masdevallia, he may have never seen these little wonders. He did see some even stranger orchid flowers but failed to understand the depths of their deceit. Investigating orchids of his local area in Kent, he documented many cases of food deception, but he puzzled over plants in the genus *Ophrys*. He showed that the bee orchid, *Ophrys apifera*, was almost completely self-pollinating, but that other species such as the fly orchid, *Ophrys insectifera*, were only pollinated rarely; he was baffled by how these species persisted in nature. In a footnote he repeated a fact from an 1829 Kentish flora: 'Mr. Gerard E. Smith, in his Catalogue of Plants of S. Kent, 1829, p. 25, says: "Mr. Price has frequently witnessed attacks made upon the Bee Orchis by a bee, similar to those of the troublesome Apis muscorum." What this sentence means I cannot conjecture.' It boggled the imagination that a bee would attack an orchid, when so often they served as pollinators, carrying pollinia from one plant to another.

However, *Ophrys* flowers are indeed 'attacked' by insects. More than 50 years after Darwin's book was published, the phenomenon was investigated in detail by a colonial judge in Algiers, Maurice-Alexandre Pouyanne, whose interest was piqued by a Swiss botanist named Henry Correvon, with whom he corresponded. In Algeria Pouyanne observed insects that were visiting *Ophrys speculum* (the mirror orchid) behaving most strangely – 'its abdomen dives at the bottom in the long red hairs that look like the labellum's [lip's] bearded crown. The abdomen tip becomes agitated against these hairs with disorderly, almost convulsive movements.' These were not female bees mistakenly looking for nectar though, but male scolid wasps (*Dasyscolia ciliata*). It looked like the insects were trying to copulate with the flowers. A few years later in southern France the same phenomenon was observed by British naturalist Colonel Masters John Godfery, and almost certainly by his wife Hilda whose involvement in the discovery is never acknowledged, 'by watching cut flowers in vases on hotel verandas in the Mediterranean region'.

ABOVE: Even to human eyes, the flowers of the bee orchid, *Ophrys apifera*, look for all the world like insects having a rest from flight perched along the stem.

It took a female amateur naturalist in Australia, however, to really unravel this apparent mystery. Edith Coleman was an Englishwoman whose family emigrated to Melbourne when she was 13 years old. She had a passion for the outdoors and went on to become a schoolteacher in rural Victoria, until her marriage, when she was barred from teaching under the conventions of the time. She and her husband, a motor car enthusiast when automobiles were just becoming fashionable in Australia, lived on the outskirts of Melbourne, where she raised her children and immersed herself in the flora and fauna of the Australian bush. Coleman was no dilettante, though; she took nature and its ways seriously and became a prolific nature writer on many subjects. She was fascinated by orchids: 'Surely there is no more fascinating hobby than studying orchids. ... With me the love of these shy blooms is not an isolated attachment.' She worked with the tiny grassland and forest orchids of southern Australia – they grew all around in the outskirts of Melbourne and near Goongarie, the Coleman's country retreat. She used her writing to engage people with nature – her first popular orchid article was entitled 'Pucks and Ariels of the forest'. Her articles were not only appreciated by the readers of women's magazines, newspapers and natural history newsletters, the pre-eminent orchid biologist of the day, Oakes Ames of Harvard University, wrote to her, saying 'Your papers are most welcome and I want all of them, including the newspaper clippings.' Her knowledge of orchids was such that she discovered several species, including the graceful leek orchid; the scientific name of this species is *Prasophyllum pyriforme* E.Coleman – the 'E.Coleman' after the genus and species name means she gave this plant its name as a new species. Coleman was the first woman to be awarded the Australian Natural History Medallion by the Field Naturalist's Club of Victoria. She wrote popular field guides to the wattles of Australia, so it is not entirely fair to say she was saved from complete obscurity by her discovery of pseudocopulation, as pollination by sexual deception came to be known.

With Ethel Eaves, Coleman pioneered the use of photography in natural history observation and documented the strange behaviour of wasps on flowers of the small-tongue orchid, *Cryptostylis leptochila*, in a paper published in a natural history journal called *The Victorian Naturalist*. In it she appealed for entomological help in working out the phenomenon, observed by her daughter, of wasps entering with their abdomens towards the centre of the flower, rather than head-first. Clearly no real help was forthcoming, because later on that year she published a follow-up study in which she conclusively showed that the male wasps were seeking out the orchids using their scent – she covered up the flowers with cloth and the wasps still arrived – and that the pollinia attached to the abdomen of a wasp visiting one flower were transported to another plant, thus effecting cross-pollination. 'It is, I think, safe to assume that, as the orchid *C. leptochila* is visited by male wasps only, and as these are seeking neither nectar nor edible tissue, they are answering an irresistible sex instinct'. She was sure the wasps deposited sperm on the flowers, thus clinching the argument, but in her usual manner was modest about her discovery – 'although I am confident I have spermatozoa on the slides they are very difficult to distinguish from vegetable matter, pollen grains, etc.' Insects were trying to mate with flowers – 'based on the resemblance of the flower to a female wasp. Even to our eyes, the likeness is apparent. To the inferior eyesight of the insect, the resemblance may be still more convincing.' More recent examination with high magnification microscopy has shown Coleman was indeed right – male wasps do ejaculate in the flowers.

Edith Coleman went on to study the phenomenon of pseudocopulation in many other species of local orchid. She was in the hotspot of orchid sexual deception – of the known sexually deceptive orchid genera, the majority are Australian. The ichneumon wasps Coleman observed on *Cryptostylis leptochila* have been given the common name of 'orchid dupe wasps' (*Lissopimpla excelsa*) and exclusively pollinate all species of *Cryptostylis*. Some wasps back into the flowers in a similar way to *Lissopimpla*, others are trapped by semi-closed flowers and still others are trapped by the plants via extraordinary hinge mechanisms. The Australian hammer orchids (*Drakaea*) have a labellum that is modified into an enlarged blob that looks like a flightless female thynnid wasp mounted on a hinged stalk. When the male wasp approaches and attempts to fly away with the pseudo-female, as he would with a real flightless female wasp, the hinge is triggered and the insect is flipped onto the column on his back, where the pollinia then stick, only to be deposited on the next flower with which he tries to mate.

Solitary and parasitoid wasps in a number of families are the most common pollinators of sexually deceptive orchids worldwide; we see this as unusual behaviour for normally predatory wasps because they are not often thought of as pollinators, but wasps are certainly important for these orchids. European *Ophrys* species are pollinated by various species of solitary bee, but in some South American orchids social bees exhibit this behaviour. Perhaps surprisingly, but not to those who study them, flies are also duped by orchids. Coleman thought she saw mosquitoes entering the flowers of the greenhood orchid (*Pterostylis*), but fly pollination by sexual deceit in *Pterostylis* was not confirmed until 2019, when fungus gnats of two different families were documented as pollinators of the trap flowers of Australian *Pterostylis* orchids with very different flower shapes. Like Coleman, the researchers concealed flowers and found that male flies found

them by scent alone. While in flowers, visiting flies showed stereotypical mating behaviours, such as wing fanning, abdomen curling and probing with genitalia; that they left with pollinia attached to their bodies showed that they made contact with the reproductive parts of the flower. But so little is known about the biology of these fungus gnat flies that there are still many things to find out, both about the flies themselves and the ways in which they pollinate orchids.

Sexual deception has evolved many times from food deception in orchids, but all sexually deceptive orchids studied so far attract their pollinators via volatile, airborne chemical signals; as previously mentioned, Coleman and others showed this by concealing flowers, and still the insects arrived. Little work has been done on sexually

ABOVE: In visiting the flowers of *Cryptostylis*, such as this *Cryptostylis erecta, Lissopimpla* wasps back into the flowers' abdomen first; they are clearly not seeking a nectar reward.

deceptive orchids in all places where they occur, but all species tested in both Europe and Australia emit chemicals that mimic the pheromones used by female insects to attract males – usually blends of different compounds. Orchids use sex signals from the insects themselves to lure these pollinating males to flowers. Pollinators approach the flower with a zig-zag flight path, characteristic of following a scent trail, and can be lured with synthetic baits made from the relevant compounds. The early work done with European species of *Ophrys* suggested that scent compounds used in pollinator attraction were already present in the plant and were co-opted for use in this new function; plants were pre-adapted. Recent discoveries in Australian sexually deceptive orchids, however, have shown completely new compounds to be used in pollinator attraction, suggesting evolutionary novelty also has a role to play. Whether via pre-adaptation or novelty, the orchids adapt and adjust their volatile offering to the local insect fauna; this makes sense in a system that must remain tightly specific. Chemical signals can work over long distances, especially with larger insects such as wasps, but textures are thought to play a role close up, in guiding the insect to make contact

ABOVE: In experiments with concealed and visible flowers, those hidden were equally attractive to the fungus gnat pollinators of this species, *Pterostylis nutans*.

with the orchid column in the right way for pollinia to be attached and then for them to contact the stigma in the next flower. Brushes of hairs, shiny patches, different labellum shapes, floral traps – all these contribute to orienting the insect once on the flower.

There is a very wide range of variation in how many insects are fooled by orchid flowers; in Australian orchids the percentage varies between 4 and 90 per cent of insects tricked into pollination. But insects do not remain fooled by the cunning deceivers for long. When sexually deceptive orchids are introduced to an area, the pollinators arrive quickly but soon learn the position of the plant and avoid it. Their sense of smell has not been dulled or habituated; if the plant is moved, they go for it again. Learned avoidance by pollinators after visiting sexually deceptive orchids is well documented for many species and is likely to be an evolved response to the costs imposed by the deception. And costs there can be – if the insect deposits sperm it is wasted; time is also wasted trying to mate with orchids rather than conspecific females; and sperm depletion could mean unsuccessful mating after an orchid visit. All this can lead to reduced fitness for the male insect in question, and in turn favour learned avoidance behaviour. But just how insects learn is not well understood, and neither is the neurobiology of response to chemical signals. It seems we know more about the orchid side of this plant–pollinator interaction.

What we know about sexual deception has increased a lot since Linnaeus and Darwin, but most of the work has been done with only a few European and Australian orchids – the importance, or even the incidence, of this in orchid hotspots such as South America is an almost blank slate. In 2005, botanists documented pseudocopulatory pollination in the tiny South American orchid *Lepanthes* by fungus gnats that themselves turned out to be a new species. So much is unknown, but the story of Edith Coleman shows you don't have to be a professional scientist to make stunning contributions to the study of nature – all it takes is patience, attention to detail and a love of the subject. Nature was part of Edith Coleman's life, and if you let it be part of yours too, who knows what you will discover!

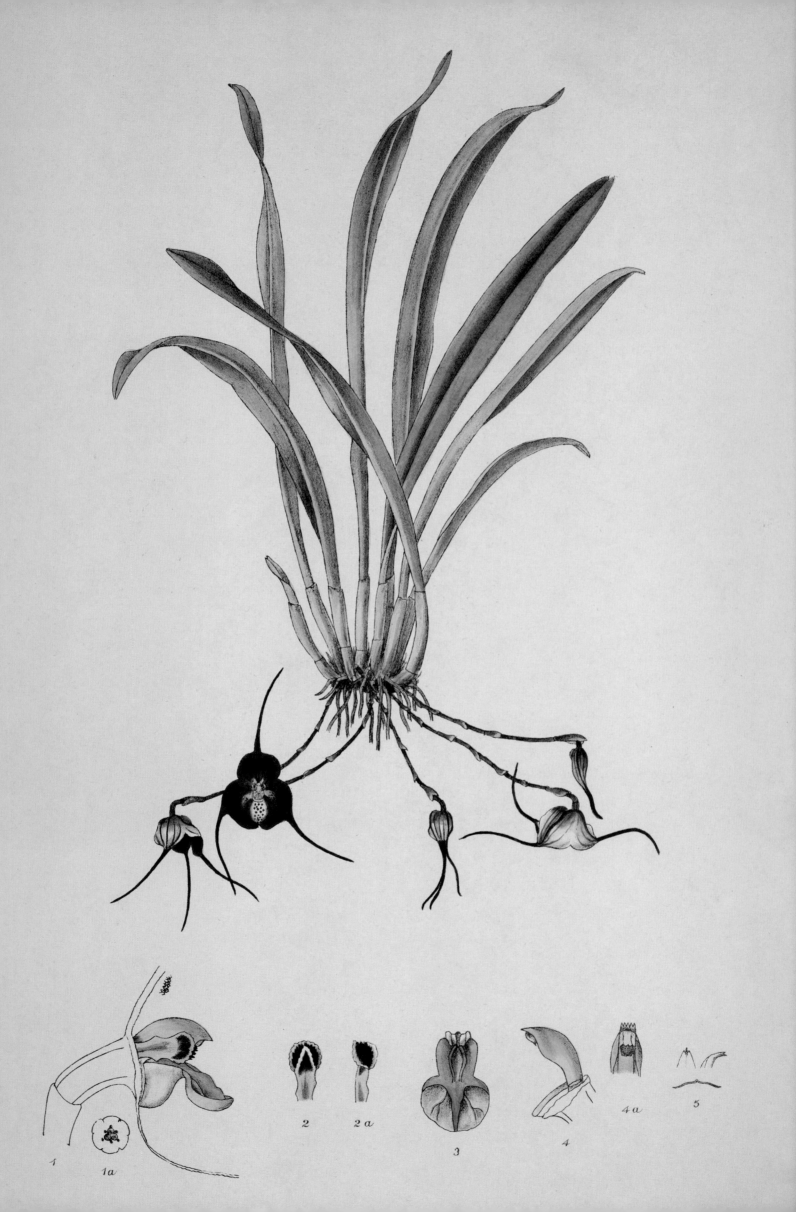

1   1a   2   2a   3   4   4a   5

MASDEVALLIA CHIMÆRA Rchb. f.

ABOVE: *Dracula chimaera*. Florence Woolward was a self-taught artist whose considerable talents were spotted by the 9th Marquess of Lothian (Schomberg Kerr, a diplomat and politician); he employed her to illustrate his collection of orchids that was said to rival that at the Royal Botanic Gardens at Kew. Her illustrations were accompanied by descriptions that were not attributed to her at first, as was usual for women in the nineteenth century. A reviewer in 1892 celebrated the rectification of this omission: 'Miss Woolward, who is responsible for the whole of the work, has overcome the unfortunate modesty which did not previously allow of her name appearing on the cover...'

OPPOSITE: *Dracula benedicti*. One of the names for this orchid had the species epithet 'troglodytes' meaning cave-dweller. The original description marvelled, 'Étrange et admirable petite plante! La fleur vient se place sous une arcade de feuillage...; on dirait une petite grotte, de coleur brune et sombre, au fund de laquelle on apperçoit une petite figure sculptée ressemblant á un gnome... Cette figure nos a fait penser á un Troglodyte cache dans une caverne....' (Strange and admirable little plant! The flower is placed under an arcade of foliage ...; it looks like a small cave, brown and dark in colour, at the bottom of which we can see a small sculpted figure resembling a gnome ... This figure made us think of a Cave-dweller hidden in a cave.)

MASDEVALLIA CHIMÆRA *Rchb.f.var* ROEZLII.

ABOVE: In Florence Woolward's painting of this orchid, *Dracula roezlii*, originally described as a variety of *Dracula chimaera*, you can clearly see the extraordinary resemblance of the labellum to the underside of a mushroom complete with gills. The pale pinkish colour contrasts greatly with the dark purplish black sepals, and along with the scent was a certain draw to fungus flies looking for a place to oviposit.

OPPOSITE: *Ophrys apifera*. Writing to his friend and mentor Charles Lyell on October 1st, 1861, Charles Darwin revealed that he too could be frustrated by orchids: 'But I am very poorly to-day, and very stupid, and hate everybody and everything. One lives only to make blunders. I am going to write a little book for Murray on Orchids, and to-day I hate them worse than everything. So farewell, in a sweet frame of mind.' Bee orchids vexed him greatly.

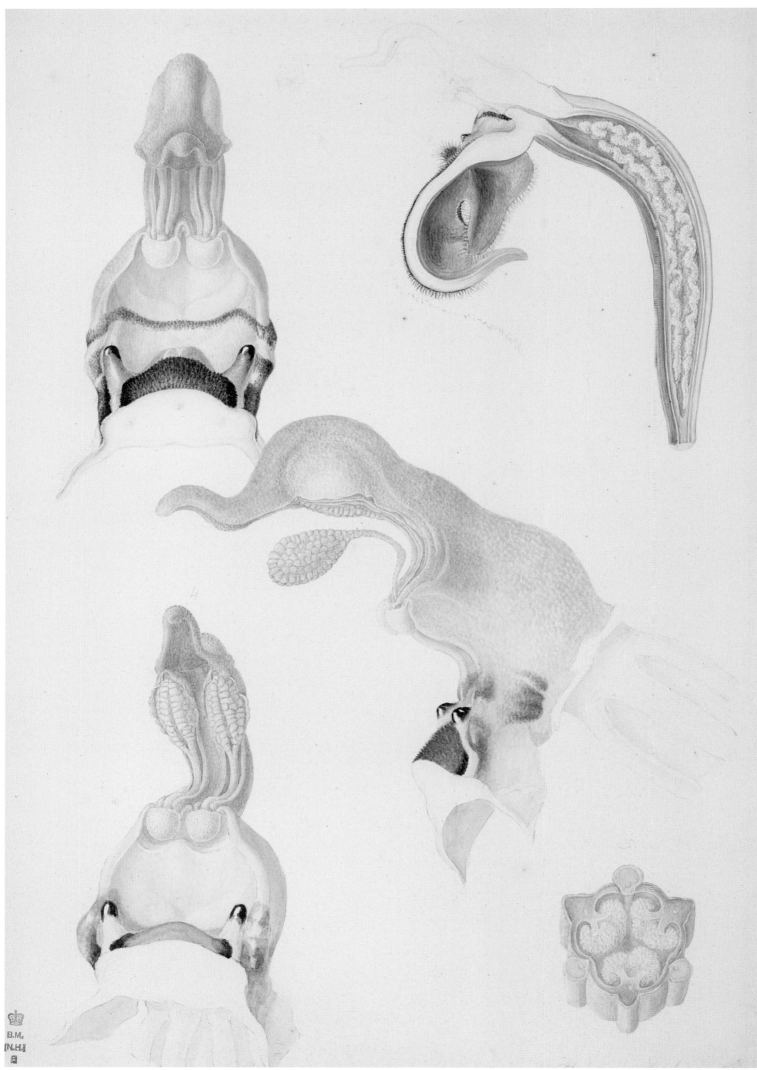

4

February 4th 1834, Kew garden
from Trinidad —

Natural size.

X 3.

X 10.

5

ABOVE: The flowers of the North American grass pink, *Calopogon tuberosus*, are non-resupinate, that is, at flowering time the labellum is uppermost in the flower. The grass pink is unusual in that the ovary does not twist; in some other non-resupinate orchid flowers the ovary twists through 360° to end up with the labellum uppermost when the flower opens.

OPPOSITE: Flies have tetrachromatic vision, they perceive colour in four categories – ultraviolet (UV), blue, yellow and purple. They see the tiny flowers of these miniature orchids, *Trichosalpinx orbicularis*, in very different ways than humans do. The almost transparent sepals of the tiny flower drawn here by Franz Bauer contrast with the darker lip to our eyes, but to a fly, what might this look like?

ABOVE: The labellum of the grass pink, has an elaborate yellow-tipped fringe at its apex. These protuberances are thought to mimic the anthers of other co-occuring flowers, when bees visit to harvest pollen their weight causes the hinged labellum to suddenly bend downwards, depositing the bee on the column where the pollinaria are attached to their backs. After each visit, the labellum springs back to its original position. Only bees of the right size can make the hinge bend, too small and nothing happens, too big and the labellum bends the wrong way. In this painting, Franz Bauer has positioned the flowers the wrong way up; in nature the fringed labellum is uppermost.

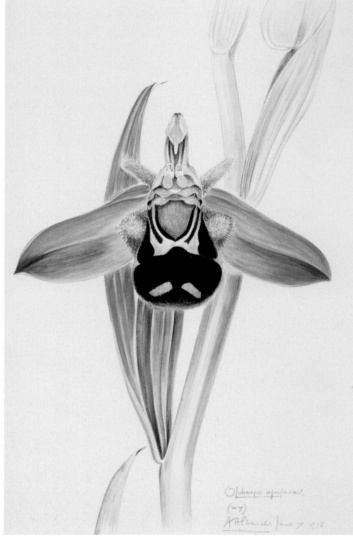

ABOVE: The fleshy protuberances at the sides, and the yellowish appendix at the bottom, on late spider - orchid, *Ophrys fuciflora*, flowers may provide tactile stimulation to insects during pseudocopulation, as does the texture of the labellum itself. These cues would function once the insect is close to or on the flower, complementing and reinforcing the fragrance cues that are perceived from far away.

ABOVE: In this detailed study of the bee orchid flower, *Ophrys apifera*, the pollinaria appear to be dangling into the curve of the column. This drops the pollen masses directly onto the stigma, self-pollinating the flower. The high fruit set Charles Darwin observed in bee orchids is because they are mostly self-pollinated in Britain.

OPPOSITE: Franz Bauer sketched the bee orchid flower, *Ophrys apifera*, from all angles, demonstrating all its parts in exquisite detail. The mechanism by which the pollinaria are attached to the heads of male long-horned bees (*Eucera*), the pollinators of bee orchids in continental Europe, is easy to see – the yellow pollinia are on long, flexible stalks (caudicles) with the sticky viscidium at their base. When the bee enters the flower, he touches the sticky area and the whole unit lifts off. As it dries, the caudicles contract, positioning the pollinia for deposition in the stigmatic opening when the bee tries to mate with another flower.

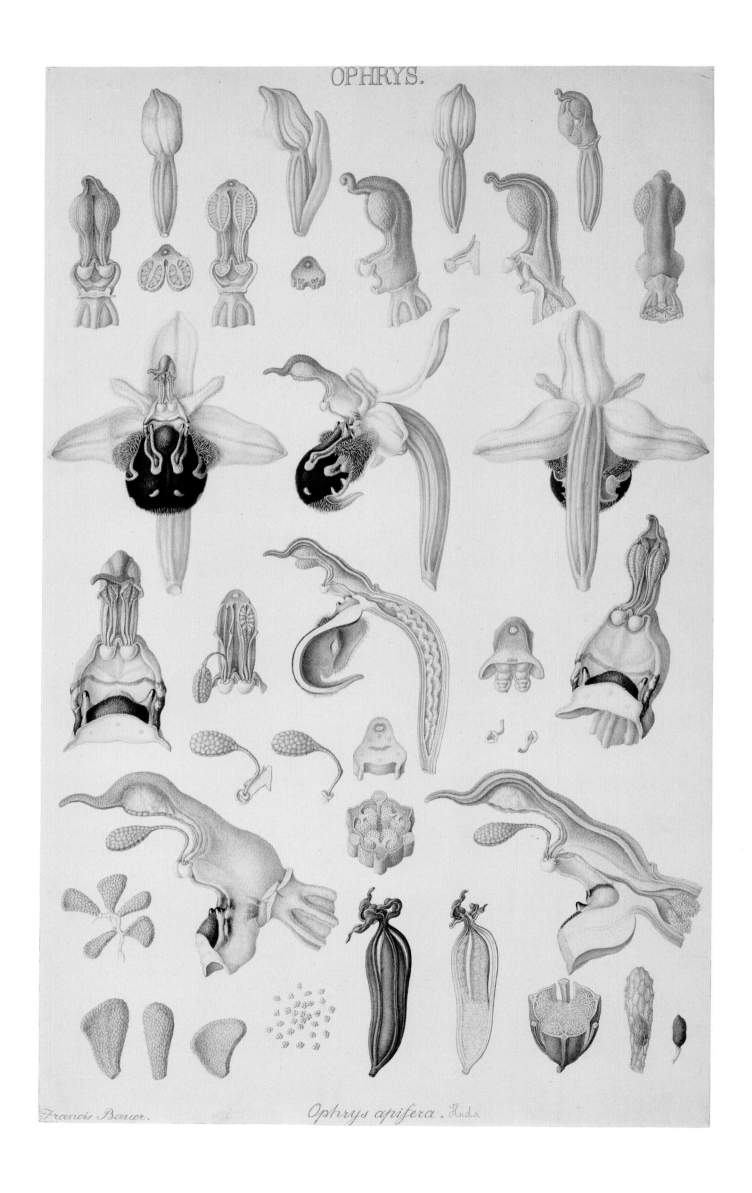

Francis Bauer.

Ophrys apifera. Huds.

ABOVE: The labellum of the fly orchid, *Ophrys insectifera*, is dark coloured with two glossy depressions that look like eyes near the column and a blue, somewhat iridescent patch in the centre – these were thought to mimic the eyes and wings of a female insect. The fly orchid is not pollinated by flies, but rather by male wasps of the genus *Argogorytes*. These solitary black wasps with a few yellow stripes are about a centimetre long and nest in dry banks; the females never visit orchids.

OPPOSITE: The red rustyhood, *Pterostylis rufa*, has translucent panels on the hood that is formed by the fusion of the dorsal sepal and two lateral petals. It is thought that this creates a window that allows the tiny fungus gnats, trapped in the flower by the hinged labellum, to find a way out, hopefully with pollinia attached. Robert Brown first described the genus *Pterostylis* in 1810 after he had collected a number of different types of these little plants during his voyage with HMS *Investigator*; his specimens are in the herbarium of the Natural History Museum, London.

ABOVE: *Calochilus campestris*. Beard orchids are pollinated by male scolid wasps of the genus *Campsomeris*. It is thought that each species of beard orchid attracts only a single species of wasp. The wasps enter the flowers headfirst, orienting themselves by the reddish-purple fringed beards. In their vigorous movements trying to mate with the flowers they push themselves into the column and break up the pollinia, thus smearing pollen on both their heads and on the stigma.

ABOVE: *Corybas unguiculatus*. Helmet orchids were thought to trick their presumed fly pollinators into laying eggs in the flowers – brood site deceivers. Recent work in the Australian state of Victoria shows that this is not the case. These tiny flowers produce a substance, at the base of the column, that is fed upon by both males and females of fungus gnats just after they emerge from their pupae in rotting logs near the helmet orchid populations.

OPPOSITE: The labellum of this species of tropical African *Bulbophyllum* is fringed with delicate dark reddish-purple hairs that move constantly in the slightest breeze. These are thought to lure the dipteran pollinators of these tiny flowers by imitating a fly mating swarm. In some *Bulbophyllum* species the antenna-like or horn-like structures on the column emit an attractant odour.

Plate 4.

1.

2.

3.

4.

5.

ABOVE: The three large and showy sepals of *Masdevallia* are similar to those of *Dracula*, but the bright red flowers of *Masdevallia ignea*, or banderita, suggest it might be pollinated by birds like hummingbirds that seek rich nectar rewards. But no, in Colombia these flowers are visited and presumably pollinated by female drosophilid fruit flies that 'performed different activities on the flowers'. In Arthur Harry Church's cross-section you can see the tiny hinged labellum, and the lack of a spur or nectary holding nectar for a reward.

ABOVE: *Prasophyllum parvifolium*. Another species of leek orchid, *Prasophyllum colemaniarum*, was named for Edith Coleman and her daughter Dorothy but is now considered a synonym of *Prasophyllum odoratum*. Both these species were described by Richard S. Rogers, who, though sceptical of Edith's work on pseudocopulation, soon became convinced and a lifelong friend. *Prasophyllum* is one of the few honest genera in its tribe; flowers of leek orchids offer nectar as a reward to their pollinators. A wide variety of insects have been observed visiting these flowers – bees, wasps, hover flies and even the occasional beetle.

ABOVE: The flowers of *Diuris maculata* are very similar in colour and shape to 'eggs-and-bacon' species of the pea family growing in the same area; they are food-deceptive mimics, these orchid flowers offer no reward. Some of these flowers are similar in colour to us, but the orchids are even better mimics to the bees, whose trichromatic vision has peaks that extend from blue and green into the ultraviolet. The orchids have a UV absorbent sham nectar guide that makes the flowers look even more similar to a bee's eyes than to ours.

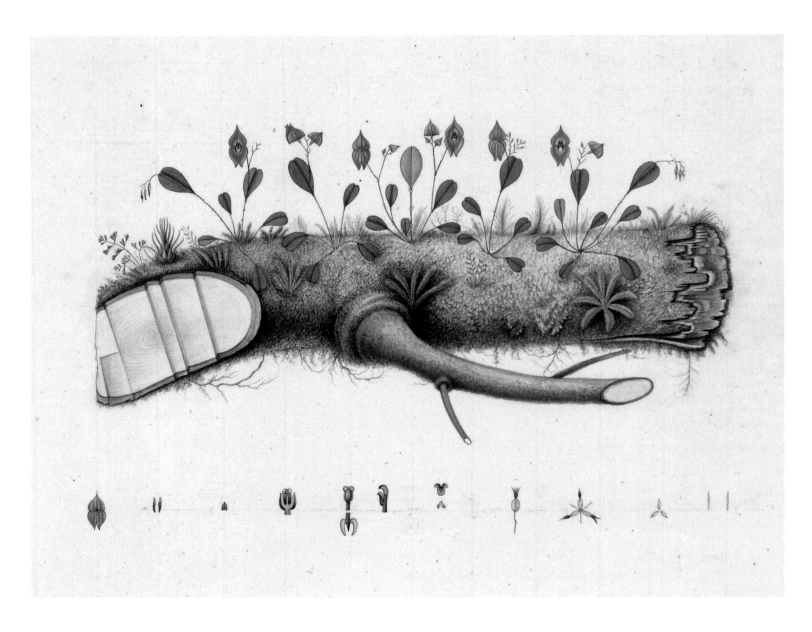

ABOVE: This diminutive orchid, *Lepanthes matisii*, is named for Francisco Javier Matis Mahecha, the painter who accompanied the expedition sent by King Charles III of Spain to explore 'New Granada' – today's Colombia (1783–1816). The Expedición Botánica al Virreinato de Nueva Granada was led by José Celestino Mutis and ranged widely over the Spanish world of northern South America. The results were not published in Mutis's lifetime, but began in the twentieth century, a joint venture between the governments of Spain and Colombia. Matis Mahecha clearly depicted the long-tailed labellum and deeply lobed anther cap that are key characters in this group of sexually deceptive epiphytic orchids.

OPPOSITE: Orchidologists have long suggested that tiny flies pollinate the equally tiny flowers of pleurothallid orchids, like this *Pleurothallis bivalvis*, but observing any behaviour is challenging – making observations is tricky with such tiny subjects. Flies do indeed pollinate species of this group that have been investigated, in a variety of ways, both at night and in the day. Some collect nectar, others lay eggs in the flowers, others use flowers as places to find mates – the diversity is amazing.

RIGHT: *Ophrys insectifera*. In the mid-twentieth century the Swedish entomologist Bertil Kullenberg suggested that 'the [female] abdominal scent has the same irresistible attraction and exciting effect for males on copulation flight as the scents of *Ophrys* flowers.' He extracted compounds from female insect bodies and from orchid flowers and experimented with attraction of male bees and wasps. He was ahead of his time. These studies preceded the discovery of the first insect sex attractant (in silkworm moths) by several years.

OPPOSITE: *Bulbophyllum* is the largest genus of flowering plants on the hyper-biodiverse island of New Guinea with more than 650 species, most of them growing only there. The flowers are remarkably diverse, with a plethora of strange and wonderful forms. Franz Bauer's painting of *Bulbophyllum hirtum* was probably done from rehydrated flowers. In life the flowers of this species are pale cream and yellow, not brown. The image shows the characteristic *Bulbophyllum* hinged labellum, which if landed on would catapult an insect straight into the column.

RIGHT: The early spider-orchid, *Ophrys sphegodes*, has nothing to do with spiders but is pollinated by male miner bees (*Andrena*) who emerge in the spring before the females. In their search for mates they mistake the orchids for female bees and attempt to mate with them. It is common in Europe and considered rare in Britain, but happily colonized the heaps of chalk dumped after Channel Tunnel construction.

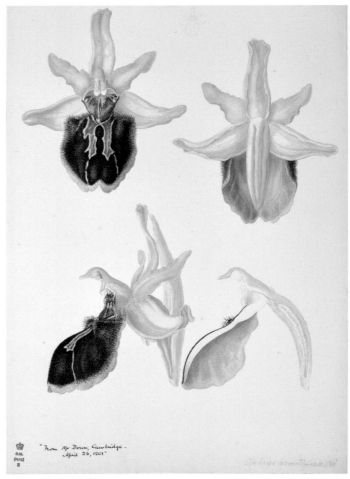

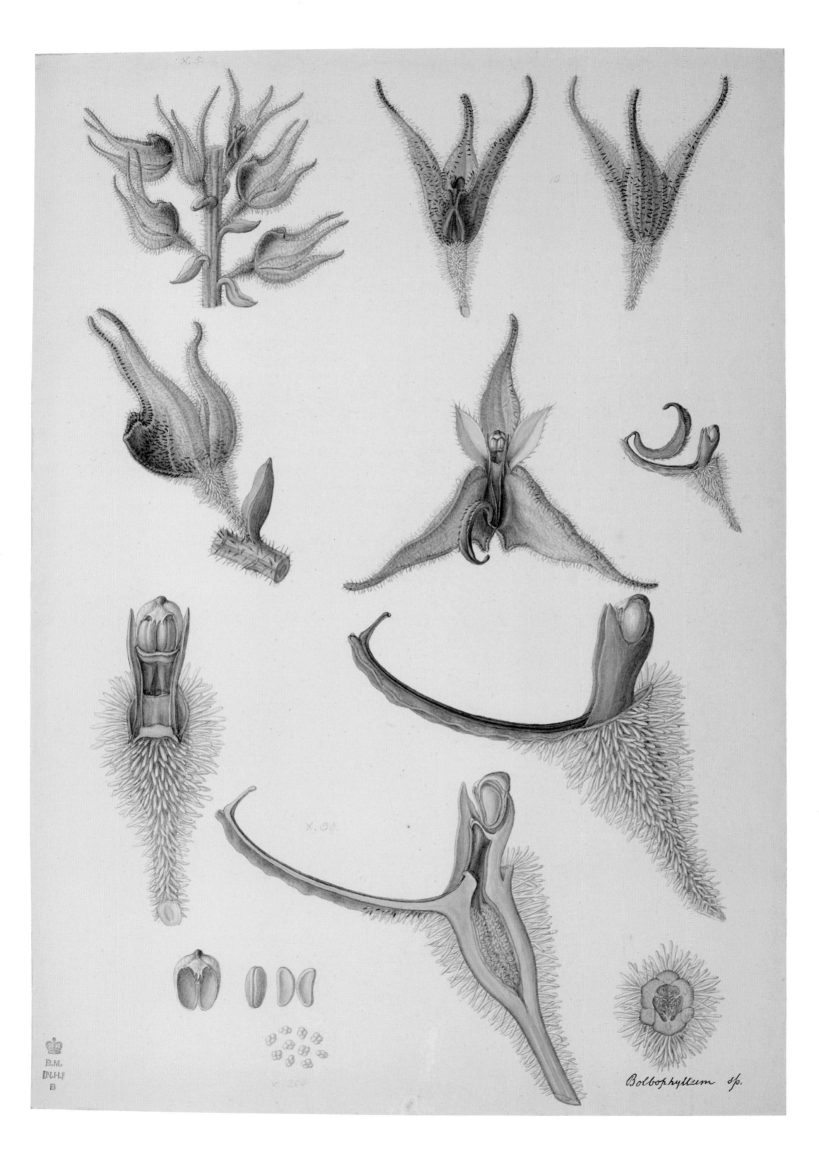

Bolbophyllum sp.

RIGHT: Flying male thynnine wasps mate with the wingless females, who climb onto bare stems and attract males with scent. *Chiloglottis* orchids, like this *Chiloglottis reflexa*, trick these wasps by both emitting similar odours from the dark protuberances (calluses) on the labellum that look like wingless females – at least to the deluded males. As the male wasp grasps the callus the labellum collapses against the hooded sepal, trapping the insect, who then escapes by crawling forward out of the trap through a gauntlet of stigma, viscid secretions and pollen-containing anthers.

OPPOSITE: The shield-like structure in the centre of this flower is a modified sterile anther, or staminode. It has an important function in the pollination of some slipper orchids. When an insect tries to land on the staminode using the central knob as a perch, it slips and falls into the bucket-shaped labellum, from where it must exit by crawling up a ladder of hairs, then squeeze out at the base of the column to emerge from a tiny opening, getting the sticky pollen all over its back. The staminode is thought to be a generalized food mimic. Many *Paphiopedilum* species, like this *Paphiopedilum insigne*, are pollinated by pollen-collecting hover flies (Syrphidae), who must consider a huge yellow object an irresistible treat.

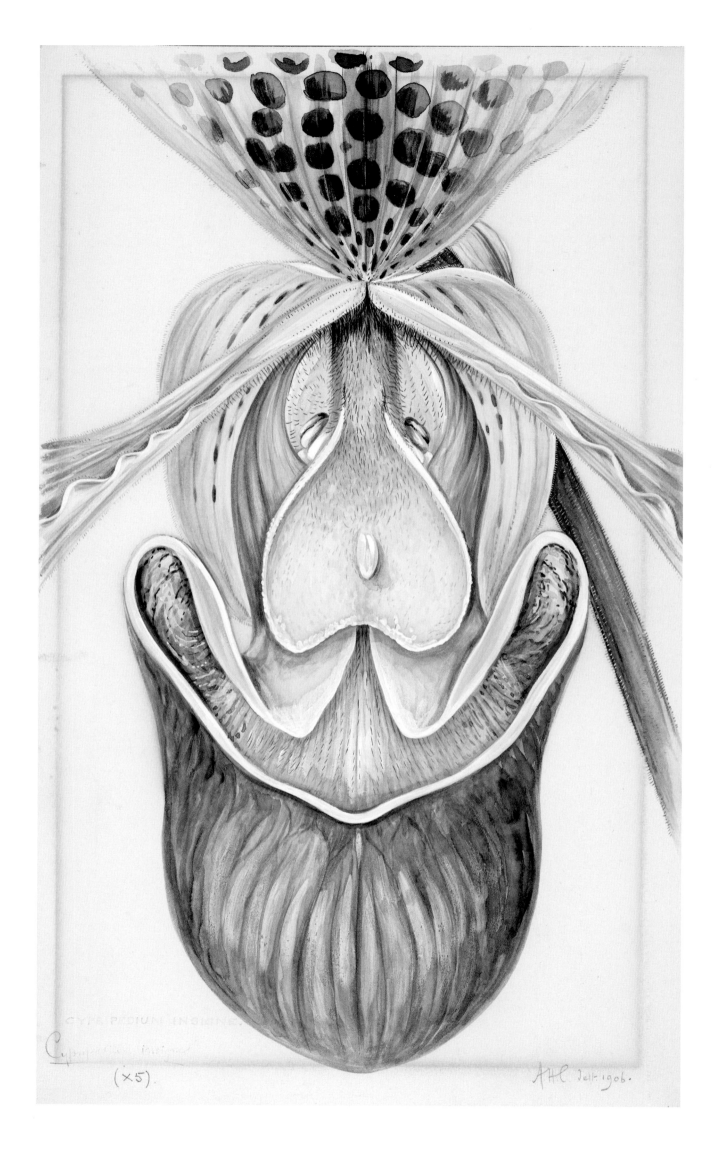

CYPRIPEDIUM INSIGNE.

Cypripedium insigne.

(×5).

A.H.C. delt. 1906.

Postscript

We live on a changing planet. To be fair, it has always been changing – change is an integral part of Earth's history over the millennia, but today, in the Anthropocene, an era defined by human impact, the rate of change is faster than ever. Plants are under threat today as never before. Their habitats are shrinking and disappearing. The expansion of land for agriculture to feed the growing human population involves destruction of complex habitats. And, a changing climate means that extreme events like droughts and wildfires are ever more common. Deforestation harms more than just trees; it is estimated that about one-fifth of all plant species are at risk of extinction.

The tropical forests of the world harbour more than half of all plant species, and among these are the many tropical orchids. These denizens of the forest are often rare, usually out of sight; they are always at risk when forests are felled, altering beyond recognition the unique habitats of canopy and understorey. As we have seen in this book, orchids have many adaptations to deal with the specialized extreme habitat of the canopy, but the growing conditions that result when forests disappear defy even them. It is not only tropical epiphytic orchids that are at threat. The myriad other organisms that interact with orchids over the course of their lives – the insects that pollinate them, the unseen fungi that help the tiny orchid seeds along at the beginning of their lives – these species too are threatened by widespread environmental change. The orchids of other habitats are not safe either. The temperate grasslands, where many orchids thrive, are among the most threatened biomes on Earth; they are often seen as ideal for agriculture and human development. Just as we often fail to notice plants as integral parts of wildlife (a trait known as 'plant blindness' or 'plant awareness deficit') we often fail to connect the loss of biodiversity far away – orchids included – with our own everyday actions. Human impacts, from deforestation to climate change, have reached into every corner of the globe. Cutting down tropical trees destroys the trees upon which perhaps unknown epiphytic orchids grow, dooming species not yet described to extinction; the use of fossil fuels has changed the climate such that some temperate orchids are blooming earlier than ever, out of synch with their pollinators. This means that they are not setting seed to make that vital next generation.

Many orchid species are rare in nature. A rare orchid protected in a national park, or the subject of a national conservation strategy, has a chance, but the vast majority of plant species are not so lucky; there are many that will, as Alfred Russel Wallace, one of my natural history heroes, eloquently said in 1862, 'perish irrecoverably from the face of the earth, uncared for and unknown.'

When we hear about the trade in endangered wildlife it is usually large mammals - rhinos, elephants, pangolins – that spring to mind. Illegal trade in wild plants is also a thriving business. At the heart of this are orchids. Because of their beauty and rarity, these plants spurred an orchidomania in the nineteenth century that continues today. Whether the commercial collector, who strips a hillside of last wild population of a rare slipper orchid, or the casual collector, who brings back an orchid from a holiday because 'it's only one and no one will ever miss it if the rainforest is being cut down anyway', any person's export of orchids without permission is illegal. It is also highly lucrative; fanatical collectors will pay a great deal of money for something no one else has, whether it be an original Picasso or a rare orchid.

The Convention on International Trade in Endangered Species, CITES, regulates the commercial trade in wild plants and animals that might be threatened. Absolutely all orchid species are listed as protected by CITES; over 70% of all species, not just plants, listed are orchids. Advances made in artificially propagating orchids mean that most of the formal, legal, orchid trade comes from artificially propagated plants, not from the wild. Your supermarket *Phalaenopsis* or prom-corsage *Cattleya* has been grown this way, as were most of the 650 million live plants traded legally between 2006 and 2015. The legal trade in orchids has its own issues; worldwide transport of plants has resulted in rapid and universal spread of serious viral orchid diseases, threatening not only the horticulture industry, but also plants in the wild. Behind closed doors though, the illegal trade in orchids still thrives, in part because the CITES regulations are poorly enforced. Little monitoring of trends in illegal orchid trade is carried out, unlike for rhino horn or elephant ivory. In 2017 the International Union for Conservation of Nature's Orchid Specialist Group recommended that, in addition to habitat conservation, effort should be put into

ABOVE: The Portuguese Jesuit missionary João de Loureiro described this orchid from what was then called Cochinchina (Vietnam) in 1790, and it has been in cultivation ever since. The striking bright red flowers of *Renanthera coccinea* are unusual in orchids.

understanding the dynamics of the illegal trade in wild plants to better address these issues. Today though the illegal orchid market is still thriving through largely undetected sales over the internet, to the detriment of wild orchids and their habitats. Globalization of trade, legal and illegal, coupled with large-scale environmental change, means effects on biodiversity can be rapid and devastating.

So, if you have just begun to love orchids, like me, or are already afflicted with orchidomania, it is important that we think carefully about how our actions, however small, can change the world for these charismatic plants and their natural environments. Orchids enrich our lives, not only with their beauty and charm, but by being integral parts of complex ecosystems that regulate the health of the planet. Our own future is intimately tied up with theirs; it is in our best interest to work for a world where the diversity of nature is celebrated, protected and where even the tiniest things are valued. Again, Alfred Russel Wallace put it best: 'If this is not done, future ages will certainly look back upon us as a people so immersed in the pursuit of wealth as to be blind to higher considerations. They will charge us with culpably allowing the destruction of some of [that] which we had it in our power to preserve.'

Bibliography

The 'go-to' volumes for anyone interested in details of orchid anatomy, morphology, relationships, ecology and taxonomy are the six lavish tomes of *Genera Orchidacearum* edited by Alec Pridgeon, Phil Cribb, Mark Chase and Finn Rasmussen – a compilation of the then-current state of knowledge about all things orchidaceous. But understanding of orchid taxonomy and relationships is constantly evolving as new knowledge and techniques are used (e.g. Karremans et al. 2016); that is what makes science exciting.

Ames, O. (1922), Observations on the capacity of orchids to survive in the struggle for existence. *Orchid Review*, 30: 229–234.

Ames, O. (1946), The evolution of the orchid flower. *American Orchid Society Bulletin*, 14: 355–360.

Arditti, J. (1984), A history of orchid hybridization, seed germination and tissue culture. *Botanical Journal of the Linnean Society*, 89: 359–361.

Arditti, J. and Ghani, A.K.A. (2000), Numerical and physical properties of orchid seeds and their biological implications. *New Phytologist*, 145: 367–421.

Arditti, J. et al. (2012), 'Good Heavens what insect can suck it' – Charles Darwin, *Angraecum sesquipedale* and *Xanthopan morganii praedicta*. *Botanical Journal of the Linnean Society*, 169: 403–432.

Arriaga-Osnaya, B.J. et al. (2017), Are body size and volatile blends honest signals in orchid bees? *Ecology and Evolution*, 7: 3037–3045.

Åsberg, M. and Stearn, W.T. (1973), Linnaeus' Öland and Gotland journey. *Biological Journal of the Linnean Society*, 5: 1–107.

Ayasse, M. et al. (2011), Chemical ecology and pollinator driven speciation in sexually deceptive orchids. *Phytochemistry*, 72: 1667–1677.

Ayasse, M. and Dötterl, S. (2014), The role of preadaptations or evolutionary novelties for the evolution of sexually deceptive orchids. *New Phytologist*, 203: 710–712.

Bateman, J. (1843), *The Orchidaceae of Mexico and Guatemala*. Ridgeway and Sons, London.

Bateman, J. (1867), *A Second Century of Orchidaceous Plants, selected from subjects published in* 'Curtis's Botanical Magazine' *since the issue of the 'first century'*. Reeve and Co., London.

Bateman, J. (1874), *A Monograph of Odontoglossum*. Reeve and Co., London.

Bateman, R.M. (2009), Evolutionary classification of European orchids: the crucial importance of maximizing explicit evidence and minimising authoritarian speculation. *Journal Europäischer Orchideen*, 41: 243–318.

Beekman, E.M. (2003), *Rumphius' Orchids: orchid texts from The Ambonese Herbal*. Yale University Press, New Haven and London.

Benzing, D.H. and Atwood, J.T. (1984), Orchidaceae: ancestral habitats and current status in forest canopies. *Systematic Botany*, 9: 155–165.

Berswerden, L. (2018), *The Orchid Hunter: a young botanist's search for happiness*. Short Books Ltd., London.

Bidartondo, M.I. (2005), The evolutionary ecology of myco-heterotrophy. *New Phytologist*, 167: 335–352.

Blanco, M.A. and Barboza, G. (2005), Pseudocopulatory pollination in *Lepanthes* (Orchidaceae: Pleurothallidinae) by fungus gnats. *Annals of Botany*, 95: 763–772.

Bohman, B. et al. (2014), Discovery of pyrazines as pollinator sex pheromones and orchid semiochemicals: implications for the evolution of sexual deception. *New Phytologist*, 203: 939–952.

Borba, E.L. and Semir, J. (2001), Pollinator specificity and convergence in fly-pollinated *Pleurothallis* (Orchidaceae) species: a multiple population approach. *Annals of Botany*, 88: 75–88.

Brown, R. (1833a), On the organs and mode of fecundation in Orchideae and Asclepiadeae. *Transactions of the Linnean Society of London*, 16: 685–738.

Brown, R. (1833b), Additional observations on the mode of fecundation in Orchideae. *Transactions of the Linnean Society of London*, 16: 739–745.

Bulpitt, C.J. et al. (2007), The use of orchids in traditional Chinese medicine. *Journal of the Royal Society of Medicine*, 100 (12): 558–563.

Burns-Balogh, P. and Bernhardt, P. (1986), Floral evolution and phylogeny in the tribe Thelymitrae. *Plant Systematics and Evolution*, 159: 19–47.

Chen, P-J. and Sheen, L-Y. (2014), Gastrodiae Rhizoma (天麻tiān má): a review of biological activity and antidepressant mechanisms. *Journal of Traditional and Complementary Medicine*, 1 (1): 31–40.

Clode, D. (2018), *The Wasp and the Orchid: the remarkable life of the Australian naturalist Edith Coleman*. Picador Books, Pan Macmillan Australia, Sydney.

Cuervo Martínez, M.A. et al. (2013), Reproductive biology of *Masdevallia coccinea* and *Masdevallia ignea* in Guasca (Cundinamarca: Colombia). *Lankesteriana*, 13: 141.

Darwin, C. (1860), Fertilisation of British orchids by insect agency. *Gardeners' Chronicle and Agricultural Gazette*, 23: 528.

Darwin, C. (1862), *The Various Contrivances by which British and Foreign Orchids are fertilised by Insects and on the Good Effects of Intercrossing*. John Murray, London.

De Boer, H. et al. (2017), DNA metabarcoding of orchid-derived products reveals widespread illegal orchid trade. *Proceedings of the Royal Society, Series B, Biological Sciences*, 284: 20171182.

Dearnaley, J.W.D. et al. (2016), Structure and development of orchid mycorrhizas. In: Martin F, (ed.), *Molecular Mycorrhizal Symbiosis*. John Wiley & Sons, Hoboken, NJ, pp.63–86.

Dressler, R.L. (1981), *The Orchids: natural history and classification*. Harvard University Press, Cambridge and London.

Dressler, R.L. (1982), Biology of the orchid bees (Euglossini). *Annual Review of Ecology and Systematics*, 13: 373–394.

Dunn, J. (2018), *Orchid Summer: in search of the wildest flowers of the British Isles*. Bloomsbury Publishing, London.

Duque-Buitrago, C.A. et al. (2013), Nocturnal pollination by fungus gnats of the Colombian endemic species, *Pleurothallis marthae* (Orchidaceae: Pleurothallidinae). *Lankesteriana*, 13: 407–417.

Ecott, T. (2004), *Vanilla: travels in search of the luscious substance*. Michael Joseph of Penguin Group, London.

Edens-Meir, R. and Bernhardt, P. (eds.) (2014), *Darwin's Orchids: then and now*. Chicago University Press, Chicago.

Edens-Meier, R. et al. (2014), Pollination and floral evolution of slipper orchids (subfamily Cypripedioideae). In: Edens-Meir, R. and Bernhardt, P. (eds.), *Darwin's Orchids: then and now*. Chicago University Press, Chicago, pp 265–287.

Endersby, J. (2016), *Orchid: a cultural history*. University of Chicago Press, Chicago.

Fay, M.F. et al. (2009), Genetic diversity in *Cypripedium calceolus* (Orchidaceae) with a focus on north-western Europe, as revealed by plastid DNA length polymorphisms. *Annals of Botany*, 104: 517–525.

Fay, M.F. and Taylor, I. (2015), 801. *Cypripedium calceolus*. *Curtis's Botanical Magazine*, 32: 24–32.

Fay, M.F. and Taylor, I. (2015), 803. *Ophrys fuciflora*. *Curtis's Botanical Magazine*, 32: 42–50.

Fogell, D.J. et al. (2019), Genetic homogenisation of two major orchid viruses through global trade-based dispersal of their hosts. *Plants, People, Planet*, 1: 356–362.

Gaskett, A.C. et al. (2008), Orchid sexual deceit provokes ejaculation. *The American Naturalist*, 171: E206–E212.

Gaskett, A.C. (2011), Orchid pollination by sexual deception: pollinator perspectives. *Biological Reviews*, 86: 33–75.

Gaskett, A.C. (2014), Color and sexual deception in orchids: progress toward understanding the functions and pollinator perception of floral color. In: Edens-Meir, R. and Bernhardt, P. (eds.), *Darwin's Orchids: then and now*. Chicago University Press, Chicago, pp 291–309.

Gerard, J. (1597), *The Herball, or Generale History of Plants*. John Norton, London.

Gerard, J. (1633), *The Herball, or Generale History of Plants gathered by John Gerarde of London Master in Chirurgerie: very much enlarged and amended by Thomas Johnson, citizen and apothecarey of London*. John Norton, London.

Gesner, K. (1771), *Opera Botanica*, pars secunda. M. Seligmann, Nuremberg.

Gigord, L.D.B. et al. (2001), Negative frequency dependent selection maintains a dramatic flower color polymorphism in the rewardless orchid *Dactylorhiza sambucina* (L.) Soò. *Proceedings of the National Academy of Sciences, USA*, 98: 6253–6255.

Gravendeel, B. et al. (2004), Epiphytism and pollinator specialization: drivers for orchid diversity? *Philosophical Transactions of the Royal Society*, B 359: 1523–1535.

Grieve, M. (1931), *A Modern Herbal*. Harcourt, Brace and Co., London.

Haelterman, D.J. (2020), Trends in illegal trade of wild ornamental orchids, and possible solutions. Orchid Specialist Group, Global Trade Programme website: https://globalorchidtrade.wixsite.com/home/single-post/2020/04/24/Trends-in-illegal-trade-of-wild-ornamental-orchids-and-possible-solutions [accessed 24 April 2020]

Hinsley, A. et al. (2017), A review of the trade in orchids and its implications for conservation. *Botanical Journal of the Linnean Society*, 186: 435–455.

Hooker, W.J. (1849), *A Century of Orchidaceous Plants*. Reeve and Benham, London.

Hooker, J.D. (1868), Thunia bensoniae. Mrs. Benson's Thunia. *Curtis's Botanical Magazine*, 94: t. 5694.

Indsto, J.O. et al. (2006), Pollination of *Diuris maculata* (Orchidaceae) by male *Trichocolletes venustus* bees. *Australian Journal of Botany*, 54: 669–679.

Jarvis, C. and Cribb, P. (2009), Linnaean sources and concepts of orchids. *Annals of Botany*, 104: 365–376.

Jersáková, J. et al. (2006), Mechanisms and evolution of deceptive pollination in orchids. *Biological Review*, 81: 219–235.

Karremans, A.P. et al. (2015), Pollination of *Specklinia* by nectar-feeding *Drosophila*: the first reported case of a deceptive syndrome employing aggregation pheromones in Orchidaceae. *Annals of Botany*, 116: 437–455.

Karremans, A.P. et al. (2016), Phylogenetic reassessment of *Specklinia* and its allied genera in the Pleurothallidinae (Orchidaceae). *Phytotaxa*, 272: 1–36.

Kelley, T.M. (2012), *Clandestine Marriages: botany and romantic culture*. Johns Hopkins Press, Baltimore.

Kreziou, A. et al. (2015), Harvesting of salep orchids in north-eastern Greece continues to threaten natural populations. *Oryx*, 50: 393–396.

Kuiter, R.H. and Findlater-Smith, M.J. (2017), Initial observations on the pollination of *Corybas* (Orchidaceae) by fungus-gnats (Diptera: Sciaroidea). *Aquatic Photographics Short Paper*, 5: 1–19.

Lack, H.W. (2003), Ferdinand, Joseph und Franz Bauer: Testamente, Verlassenshaften und deren Schicksale. *Annalen der Naturhistorisches Museum in Wien*, 104B: 479–551.

Lack, H.W. (2008), *Franz Bauer: the painted record of nature*. Verlag des Naturhistorisches Museum Wien, Vienna.

Lack, H.W. (2015), *The Bauers: Joseph, Franz and Ferdinand*. Prestel Verlag & Natural History Museum, Munich & London.

Lindley, J. (1830–1838), *Illustrations of Orchidaceous Plants by Francis Bauer*. Published in four fascicles. Ridgway and Sons, London.

Lindley, J. (1840), *Sertum orchidaceum: a wreath of the most beautiful orchidaceous flowers*. Ridgway and Sons, London.

Liu, T. et al. (2015), Highly diversified fungi are associated with the achlorophyllous orchid *Gastrodia flavilabella*. *BMC Genomics*, 16: 185.

Lyons, J.C. (1853), *Remarks on the Management of Orchidaceous Plants, with a catalogue of those in the collection of J.C. Lyons, Ladiston*. Published by the author.

Mabberley, D.J. and Moore, D.T. (1999), Catalogue of the holdings in The Natural History Museum (London) of the Australian botanical drawings of Ferdinand Bauer (1760–1826) and cognate materials relating to the Investigator voyage of 1801–1805. *Bulletin of the Natural History Museum, London (Botany)*, 29: 81–226.

Mabberley, D.J. (2000), *Arthur Harry Church: the anatomy of flowers*. Merrell & Natural History Museum, London.

Mabberley, D.J. (2017), *Painting by Numbers: the life and art of Ferdinand Bauer*. New South Publishing, Sydney.

McCormick, M.K. et al. (2009), Abundance and distribution of *Corallorhiza odontorhiza* reflect variations in climate and ectomycorrhizae. *Ecological Monographs*, 79: 619–635.

Michenau, C. et al. (2010), Orthoptera, a new order of pollinator. *Annals of Botany*, 105: 355–364.

Miura, C. et al. (2018), The mycoheterotropic symbiosis between orchids and mycorrhizal fungi possesses major components shared with mutualistic plant-mycorrhizal symbioses. *Molecular Plant-Microbe Interactions*, 31: 1032–1047.

Morales-Linares, J. et al. (2018), Orchid seed removal by ants in Neotropical ant gardens. *Plant Biology (Stuttgart)*, 20 (3): 525–530.

Morren, E. (1877), Description du Masdevallia Trogolodyte, *Masdevallia troglodytes* sp. nov. *Belgique Horticole*, 27: 97–98.

Mutis, C. (1992), *Flora de la Real Expedición Botánica del Nuevo Reino de Granada*, Vols. 1 and 2. Real Jardin Botanico, Madrid.

Nadkarni, N. (1994), Diversity of species and interactions in the upper tree canopies of forest ecosystems. *American Zoologist*, 34: 70–78.

Nunes, C.E.P. et al. (2016), The dilemma of being a fragrant flower: the major volatile attracts pollinators and florivores in the euglossine-pollinated orchid *Dichaea pendula*. *Oecologia*, 182: 933–946.

Ogilvie, B.W. (2006), *The Science of Describing: natural history in Renaissance Europe*. Chicago University Press, Chicago.

Ogura-Tsujita, Y. et al. (2018), The giant mycoheterotrophic orchid *Erythorchis altissima* is associated mainly with a divergent set of wood-decaying fungi. *Molecular Ecology*, 27: 1324–1337.

Oliver, F.W. (1888), On the sensitive labellum of *Masdevallia muscosa*, Rchb.f. *Annals of Botany*, 1: 237–253.

Orlean, S. (1999), *The Orchid Thief: a true story of beauty and obsession*. William Heinemann, London.

Ornduff, R. (1984), Darwin's botany. *Taxon*, 33: 39–47.

Palomino Mondragón, P. and Theissen, G. (2009), Why are orchid flowers so diverse? Reduction of evolutionary constraints by paralogues of class B floral homeotic genes. *Annals of Botany*, 104: 583–594.

Pansarin, E.R. and Estanislau do Amaral, M.C. (2009), Reproductive biology and pollination of southeastern razilian *Stanhopea* Frost ex Hook. (Orchidaceae). *Flora*, 204: 238–249.

Parkinson, J. (1640), *Theatrum Botanicum: the theater of plants*. T. Coates, London.

Pellissier, L. et al. (2010), Generalized food-deceptive orchid species flower earlier and occur at lower altitudes than rewarding ones. *Journal of Plant Ecology*, 3: 243–250.

Pérez-Escobar, O.A. et al. (2017), Recent origin and rapid speciation of Neotropical orchids in the world's richest biodiversity hotspot. *New Phytologist*, 215: 891–905.

Pillon, Y. and Chase, M.W. (2007), Taxonomic exaggeration and its effects on orchid conservation. *Conservation Biology*, 21: 263–265.

Policha, T. et al. (2019), *Dracula* orchids exploit guilds of fungus visiting flies: new perspectives on a mushroom mimic. *Ecological Entomology*, 44: 457–470.

Pridgeon, A.M. et al. (1999), *Genera Orchidacearum. Vol. 1: Apostasioideae and Cypripedioideae*. Oxford University Press, Oxford.

Pridgeon, A.M. et al. (eds.) (2001), *Genera Orchidacearum. Vol. 2: Orchidoideae (Part 1)*. Oxford University Press, Oxford.

Pridgeon, A.M. et al. (eds.) (2003), *Genera Orchidacearum. Vol. 3: Orchidoideae (Part 2) Vanilloideae*. Oxford University Press, Oxford.

Pridgeon, A.M. et al. (eds.) (2005), *Genera Orchidacearum. Vol. 4: Epidendroideae (Part 1)*. Oxford University Press, Oxford.

Pridgeon, A.M. et al. (eds.) (2009), *Genera Orchidacearum. Vol. 5: Epidendroideae (Part 2)*. Oxford University Press, Oxford.

Pridgeon, A.M. et al. (eds.) (2014), *Genera Orchidacearum. Vol. 6: Epidendroideae (Part 3)*. Oxford University Press, Oxford.

Ramsbottom, J. (1922), *Orchid mycorrhizae*. From Charlesworth & Co. Catalogue, Haywards Heath.

Rasmussen, H. et al. (2015), Germination and seedling establishment in orchids: a complex of requirements. *Annals of Botany*, 116: 391–402.

Raven, J.A. and Edwards, D. (2001), Roots: evolutionary origins and biogeochemical significance. *Journal of Experimental Botany*, 52: 318–401.

Reiter, N. et al. (2019a), Pollination by sexual deception of fungus gnats (Keroplatidae and Mycetophilidae) in two clades of *Pterostylis* (Orchidaceae). *Botanical Journal of the Linnean Society*, 190: 101–116.

Reiter, N. et al. (2019b), Pollination by nectar-foraging thynnine wasps in the endangered *Caladenia arenaria* and *Caladenia concolor* (Orchidaceae). *Australian Journal of Botany*, 67: 490–500.

Rendle, A.B. (1892), Notices of books – *The genus* Masdevallia, issued by the Marquess of Lothian K.T., Plates and descriptions by Miss Florence H. Woolward. Part III. £1 10s. London, Porter. *Journal of Botany*, 30: 348–350.

Romero, G.A. and Nelson, C.E. (1986), Sexual dimorphism in *Catasetum* orchids: forcible pollen emplacement and male flower competition. *Science*, 232: 1538–1540.

Roubik, D.W. and Ackerman, J.D. (1987), Long-term ecology of euglossine orchid-bees (Apidae: Euglossini) in Panama. *Oecologia*, 73: 321–333.

Roubik, D.W. (2014), Orchids and neotropical pollinators since Darwin's time. In: Edens-Meir, R and Bernhardt, P (eds), *Darwin's Orchids: then and now*. Chicago University Press, Chicago, pp 229–261.

Salisbury, E.J. (1957), Nicolas Henry Ridley, 1855–1956. *Biographical Memoirs of Fellows of the Royal Society*, 3: 141–159.

Selosse, M-A. et al. (2010), Saprotrophic fungal mycorrhizal symbionts in achlorophyllous orchids: finding treasures amongst the 'molecular scraps'? *Plant Signaling & Behaviour*, 5: 349–353.

Schiestl, F.P. and Cozzolino, S. (2008), Evolution of sexual mimicry in the orchid subtribe Orchidinae: the role of preadaptations in the attraction of male bees as pollinators. *BMC Evolutionary Biology*, 8: 27.

Sheehan, T. and Sheehan, M. (1979), *Orchid Genera Illustrated*. Comstock Publishing, Cornell University Press, Ithaca.

Stewart, J. and Stearn, W.T. (1993), *The Orchid Paintings of Franz Bauer*. The Herbert Press and Natural History Museum, London.

Stearn, W.T. (1960), Two-thousand years of orchidology. In: Synge, PM (ed), *Proceedings of the Third World Orchid Conference*. Royal Horticultural Society, London, pp. 26–42.

Strona, G. et al. (2018), Small room for compromise between oil palm cultivation and primate conservation in Africa. *Proceedings of the National Academy of Sciences*, USA 115: 8811–8816.

Strullu-Derrien, C. et al. (2018), The origin and evolution of mycorrhizal symbioses: from palaeomycology to phylogenomics. *New Phytologist*, 220: 1012–1030.

Swarts, N.D. and Dixon, K.W. (2017), *Conservation methods for terrestrial orchids*. J. Ross Publishing, Plantation, Florida.

Taylor, D. (2004), Myco-heterotroph-fungus marriages – is fidelity over-rated? *New Phytologist*, 163: 217–223.

Van der Cingel, N.A. (1995), *An atlas of orchid pollination: European orchids*. A.A. Balkema, Rotterdam.

Van der Cingel, N.A. (2001), *An atlas of orchid pollination: America, Africa, Asia and Australia*. A.A. Balkema, Rotterdam.

Varey, S. [ed] (2000), *The Mexican Treasury: the writings of Dr. Francisco Hernández* [translated by Chabrán, R., Chamberlain, C.L. and Varey, S.]. Stanford University Press, Stanford.

Wallace, A.R. (1853), *A Narrative of Travels on the Amazon and Rio Negro, with an account of the native tribes and observations on the climate, geology, and natural history of the Amazon valley*. Reeve & Co., London.

Wallace, A.R. (1867), Creation by law. *Quarterly Journal of Science*, 4: 470–488.

Warner, R. et al. (1882), *Orchid Album: comprising coloured figures and descriptions of new, rare and beautiful orchidaceous plants*. Self-published by B.S. Williams, London.

Williams, N.A. et al. (1981), Floral fragrance analysis in *Anguloa*, *Lycaste* and *Mendoncella* (Orchidaceae). *Selbyana*, 5: 291–295.

Wolff, T. (1950), Pollination and fertilization of the Fly Ophrys, *Ophrys insectifera* L., in Allindelille Fredskov, Denmark. *Oikos*, 2: 20–59.

Woolward, F.H. (1896), *The genus* Masdevallia. R.H. Porter, London.

Illustration Index

Page number, current orchid scientific name, source

Acknowledgements

Many people have helped me to come to love orchids. Trudy Brannan has been patient to a fault with the delivery of the text; Andrea Hart, Ben Nathan and Paul Cooper of the Natural History Museum Library put up with my wanting to see books and illustrations again and again; Isabelle Charmantier, Head of Collections at the Linnean Society of London, sourced books and patiently retrieved them for me; Doug Holland and Ellie Flesch of the Missouri Botanical Garden kindly provided an image that had slipped through the cracks before Covid-19 lockdown; Robbie Blackhall-Miles, Mark Spencer and Mary Parlange read the text in early drafts; Richard Bateman, Lauren Gardiner, Johan and Clare Hermans, Rudolf Jenny, Alan Karremans and André Schuiteman identified some of the illustrations that completely defeated me; Mark Carine and Jacek Wajer of the Natural History Museum herbaria were willing to talk about all things Robert Brown and Joseph Banks; Mark Chase and Mike Fay gently corrected my most egregious errors, although all mistakes herein are mine alone. I am especially grateful to my family, Victor, Isabel, Alfred, Charlotte and especially Libby, for moral support, and to the staff of the Oncology Outpatient Unit at the Royal Free Hospital for their kind care and unstinting support.

Index